P9-AGS-252

b,
and ...
"letting your fingers
do the flying".
Love,
Keith

A *terra magica* BOOK

Hanns Reich

The World From Above

Introduction and text by Oto Bihalji-Merin

Other texts by Rudolf Braunburg and Klaus Völger

HILL AND WANG · NEW YORK

We thank the Deutsche Lufthansa AG for its kind assistance.

Arise, sun, moon, stars, all of you
Who move in the heavens.
I pray you, hear me : In your midst
A new life has come.
Song of the Omaha Indians

Our century's vision of reality has acquired many new aspects. Microphotography has revealed galleries of structural forms that were never examined before. Aerial pictures have added a new dimension to sight and knowledge.

Aided by the sensitive feelers of modern instruments, we can peer into a world that no one has as yet catalogued. The wide view and the motion of the flying camera eye make seeing a kinetic adventure. Unexpected perspectives, opening in all directions, deobjectify the terrestrial surface and redefine its reality.

The airplane offers the observer a broader vision and permits the investigation of sources of raw material, revolutionizing the science of surveying and cartography, helping agriculture and forestry, enriching archaeological research, and allowing insights into the composition of the earth. Vision of this sort allows us to penetrate the concealed sphere of the genesis of the world. But the urge of mankind to rise from the ground is not motivated by practical aims alone. Although authorities in all times have tried to steer scientists and discoverers into the province of immediate usefulness, the curious mind is driven by the indomitable desire to fathom the unknown, to reach untouched areas of the earth and of knowledge, and to understand the secrets of life.

The ability to fly, which is less than two centuries old, has revolutionized man's consciousness. His first flights were like excursions into the land of dream and fantasy. The astronomer Lalande, a member of the Academy of Sciences in Paris, wrote a report on the flying experiments of his contemporaries in 1779: "Men cannot possibly raise themselves into the air, much less hold themselves above the ground, with artificial wings or other contrivances. Man was created for the earth, winged creatures for the air. We should not try to violate the laws of nature."

The limits of the possible shift with the expansion of human knowledge. Two years later, in 1781, the first aerostats glided over France. On November 24, 1783, two noblemen, Pilâtre de Rozier and the Marquis d'Arlande, rose in a balloon over Paris before a select group of statesmen and scholars and an enormous crowd of people. For the first time in human history, human eyes viewed the world from above. A week later, Professor Charles, the inventor of the gas balloon, rose into the air, accompanied by a mechanic named Robert. Charles described his feelings and impressions in a letter: "Nothing in my life will ever equal my joyful sensation as I rose from the earth. It was not pleasure but true happiness. Fleeing the fearful torments of hatred and calumny, I felt that I had put all my enemies to shame by rising above them. This moral feeling was followed by another, keener one: it was the hitherto unseen majestic painting of all nature spread before us in its endlessness. Below us, the throng of 300,000 onlookers stretched out like a field; above us, the cheerful vault of heaven, unmarred by any cloud; and in the distance, the most delightful of views. 'Mon ami,' I said to Monsieur Robert, 'how great is our happiness. I don't know how the earth feels about us, but isn't heaven in our favor? What serenity, what a breath-taking scene! If only those who mocked us were here, so that I might say: "This is what you lose, you wretched beings, when you impede the progress of science."'"

Man has always longed for unknown lands, he has always been lured by the loftiest mountains, the expanse of oceans, the distant stars. "And men go and admire high mountains and wide seas and powerful torrents and the ocean and the stars and rely on them," lamented St. Augustine. This has been true in all ages, in the days of Marco Polo, Columbus, or Magellan, and the astronauts of our own time.

Hundreds of thousands of miles have been traversed. "Everything is turned topsy-turvy. By flying over the Pole in a straight path, you go northward and then southward. Every wind blows in a northerly direction, and everywhere you look is south," wrote Richard Evelyn Byrd, the first man to fly across the North Pole and then the South Pole. A camera accompanied Byrd's polar flights in May, 1926. It had taken surveyors months and years to map out an area that a camera photographed while gliding across it. Byrd reported: "Here we are still in the Ice Age. Huge glaciers cover the land. Only a scattering of mountains juts through. There are so few naked rocks that a traveler welcomes them with almost as much joy as an oasis in the Sahara. The plateaus, fifteen hundred to three thousand yards above sea level, are crowned by huge mountains. In the interior lie the high plateaus of South Victoria and the flat peak-masses of the Pole. In some places, the ice layer is estimated to be fifteen hundred yards thick. These containers, emitting wide glacial streams, are divided by mountains and form marvelous drifts of ice, like the waterfalls of rivers."

The arctic air-route has now become a normal part of world travel. Within the airplane: man!

The early flights of man in balloons had something of the grace of the late rococo. The first aerial photographs were shot from kites, rockets, and balloons. In 1858 Gaspar Félix Tournachon (who went by the pseudonym of Nadar), a writer, draughtsman, inventor, and photographer, took photographs of Paris from a balloon. During World War I pigeons carrying miniature cameras around their necks were used by information groups as flying photographers.

The possibility of viewing and photographing the world from a great height changed our knowledge and our idea of this old planet, which we had hitherto only known horizontally. In this book we have avoided the oblique picture from the first floor of man's climb because of the similarity to the view from church steeples or mountain tops. New vision has grown out of the old perspective. Our eyes are being retrained and our intellectual receptivity is being revolutionized.

The Renaissance, by rediscovering perspective, had expanded two-dimensional space, thereby establishing the possibility of three-dimensional representation. The theory of relativity went beyond the sense of three-dimensional space obtaining in classical physics. The modern aerial picture is enriching artistic vision, not by directly influencing abstract painting (a copy of reality as seen from above would not necessarily be less naturalistic than a copy of horizontal viewing); instead, we are witnessing a support of the striving and experimenting by modern painters to overcome classical perspective. "With Cézanne, and artists after him, painters abandoned the old perspective picture of reality," writes the sociologist Alfred Weber. "By doing so, they rejected something that had been carefully developed for centuries, nay, viewed generally for millennia: the means of expression of that stratum of human life which is spatially structured. In lieu of this, they are seeking new forms of expression. Which simply means that they have a new way of viewing the world and that they are trying to convey it."

The painters of futurism, Umberto Boccioni and Luigi Russolo, celebrated the dynamics of aerial flight. In their compositions they used simultaneity and motion as an expression of their sense of life. The newly developing, dynamic vision of the space-time plane fascinated artists. Such kinetic experience led to K. R. Sonderborg's speed variations. Jean Fautrier's landscapes in heavy paint applied with a palette knife seem similar to high-altitude photographs. Many of the psychograms of the American abstract expressionists convey a sense of limitless space and distance. Jackson Pollock spread his paintings on the floor rather than propping them up on an easel. He stood on the canvas and splashed dynamic streams of color about him as a gesture of pure instinct, creating spontaneous abstract compositions, flowing and freezing forms.

"The surface is the expression of everything beneath it. It is not just a visible joint; it conceals a 'deeper stratum,'" wrote Willi Baumeister about the vision and work of modern artists.

In their visions artists occasionally leave the shores of earth. Voyages of discovery to the most far-flung places of existence inspired

Max Ernst to paint the dead world of the moon, perhaps the very face of the earth as it may look millions of years from now. The earthly, idyllic landscape has soared into the cosmic footlights of the stars. The dead planet is encircled by man-made satellites. Soon a cosmic Columbus follows them; his harbinger, the artist. Man has taken possession of the atmospheric space of his terrestrial world. That phase of the history of flight which took place within the troposphere may be regarded as closed. Now humanity is organizing man's travels through the stratosphere. Everything beyond that belongs to the conquering of interplanetary space.

Whither is man hastening, higher and higher, faster and faster, in his brief flight between birth and death? Do the starry patterns of infinity contain similar beings in quest of the meaning and the destination of the journey we call our life? Perhaps there are creatures on other stars who know more about the secrets and causes of existence? If the millennia could rise into the firmament, if all the pioneers of flying were to come back—Architas of Tarrento, Leonardo da Vinci, Roger Bacon, the Montgolfier brothers, Lilienthal, or the Wright brothers—if all of them were to ask whether their dreams had come true and whether flying had enriched man's existence, what would we reply? The airplane is no independent thing; man is in command. It can enrich, it can destroy, it can liberate and enslave. It has not brought man a new home, but it has given him a new perspective of the small and great existence of his world. The darker aspect of this new picture of the world should not be neglected: although man has risen above his planet and conquered the heavens, he has also used his flying machines to create a modern hell. As emissaries of megalomania and usurpation, metal birds of destruction threaten the existence of humanity: the sower who cannot tell if he will reap; the fisherman who does not fear tempests as much as radioactive rain; the city dweller desperately scanning the sky for the dark angels of death. Will our earth turn into a myth, a parable of vain striving? The all-devouring waters of the geological eras have withdrawn. The primeval volcanoes are settling. In a world without gravity and devoid of up or down, astronauts can reorient their senses. Perhaps the cosmic aspects of space can promote unity among the nations of the earth.

But superterrestrial standards and orders cannot cancel out earthly reality. Thus, the title of this book, *The World From Above,* retains its validity. Just as certain physical processes appear different according to the standpoint of the observer, man, dwelling on the earth and yet seeking cosmic space will dialectically unite dissimilar and even antithetical extremes. An expanded reality demands that he synthesize his earthly and cosmic visions.

The earliest known description of the world from above is *The Flight of Etana* (third millenium B.C.). Near the Kurdish village of Kunyundjik, the site of Nineveh, archaeologists have excavated the library of Assurbanipal: clay tablets relate a poetic and visionary tale of a mythical flight through space.

Etana's wife is expecting a child, a boy. The birth is difficult and drawn out. Etana prays to the sun god Shamash for the medicinal herb for birth. The God refers him to the eagle whom Etana had helped long before in a fight with a serpent. Thereupon the bird carries him to the sky, where the herb grows. A long flight, beyond the earth and its realm.

The eagle says to Etana: "My friend, I will bear you to the heaven of Anu. Lay your breast upon my breast. Put your hands on the quills of my wings. Place your side against my side."

He placed his breast upon its breast, he put his hands on the quills of its wings, he leaned his side against its sides. The weight was enormous.

For two hours it bore him aloft. Then the eagle said to Etana: "Look down, my friend, upon the earth. The land has become a hut, the sea around it a courtyard."

For two more hours the eagle bore him aloft. Then it spoke to Etana: "Look down, my friend, upon the earth. The land has become a loaf of bread and the sea a bread basket."

For still two more hours the eagle bore him aloft. Then it spoke to Etana: "Look down, my friend, upon the earth. The earth has submerged, everything is ocean."

Oto Bihalji-Merin

One of the first things stewardesses do once a plane has soared into the air is to make things appear as they did before take-off. Cocktails, newspapers, television, movies, are supposed to make the passengers forget they are flying.

But those who refuse to forget that they are flying and prefer to experience everything fully, drinking their martinis or Manhattans solely to increase their pleasure, will be exposed to the fascinations of the world from above.

The photographs contained in this book are proof of these fascinations. They teach us how to see the world from above. Such instruction would be the most interesting (and cheapest!) service that an airline company could provide. Unfortunately, passengers have to acquire this on their own. The impressions that can offer themselves to them are fantastic.

Arriving at Istanbul: An extremely fine layer of clouds has stretched out over the city and under that the city lies, its colors slightly dimmer, but all its details visible—a panoramic silk-print map. The mosques, the web of bridges on the Golden Horn, the yellow boats on the Sea of Marmara, everything as unreal as a dream, uncorporeal, unattainable. But when the airplane dives into the mist, a close-up of the city bursts forth.

Flying over the desert of Lut: The same sensation as listening to a Miles Davis record. An abstract world: lines beginning somewhere, plunging into nothingness, re-emerging, and—separated from points, cubes, and lemniscates—stretching out and weaving together. The traces and signs of the desert are manifold, but this impression of death and petrification does not evoke any association with the authors of these traces.

The coast of Cyprus as seen from an altitude of twenty thousand feet: The air is as pure as Venetian glasswork. In the heat waves of the turbines, the color spectrum of the island dissolves into gray. At the black vault of the radar tower, a series of color fields begins: the shadow-striped, wrinkled green of the mountains; the bright squares of Nikosia, sun-flower yellow and straw-brown; the emerald strips of coastal water; the cement gray of two minarets in orange flames.

Cities at dusk: With their lights on, they are like fabulous animals hung with jewels. The pale confusion of the landing field brings to mind numb polypites entangled in a sunken ship; like chains of dark aquamarines, the lights of avenues entwine public squares and parks.

Flying from Rio de la Plata toward the Andes: Iridescent, as if covered with blue glaze, the surface of the water sinks, the white cubes of Montevideo and Buenos Aires sink, the last traces of civilization fade out of sight. We can see the undulations of the approaching Andes, the first cumulus clouds brilliantly shining on the horizon, and faraway black contours of granite palisades. Through the green pampas run trails of sand yellow spots, as if a painter had splashed pigment about or as if the earth had been grazed by a powerful meteor. The mountains grow and grow beyond all imagination; they line up like dinosaur's teeth, like the weapons of an army in review, like tents braving the wind. At times the summits besiege us; the airplane glides between chasms in the afternoon refulgence. The pilot knows every wrinkle, every trough, every valley. And everywhere the spongy clouds rise on the Pacific wind, the foggy fleece of the stirred-up air dissolves; and the pilot knows them, too, knows the tides of streams and drifts, their whirlwinds, their wake, their surge, knows them better than the streets of his home town. Which is why we mention that boring seven-hour flight through the monotonous, unchanging stratus clouds. As we were soaring upwards they had clustered around the window panes, obstructing our view of ocean and borders.

One of the characteristics of these ill-bred clouds is their lack of respect for the beauties of nature and the glory of antiquity. With nihilistic phlegm, they extinguish everything alike, the temples of Baalbek and the coasts of Brazil. At the height of their excitement, the clouds emitted a uniform, dispassionate drizzle. We were soaring over a whole array of interesting landscapes and huge metropolises; all we could do was to remain seated, struck blind as it were, glaring at the instruments and sulking. The only break in the monotony occurred when a passenger stepped into the cockpit because he wanted to have a glimpse of the pilot's interesting work. After waxing enthusiastic about the beauty of flying he asked how soon we would be coming to the city of P. "In three minutes," we replied. The man shook his head in disbelief: "It's a huge city," he said. "I was born there. We ought to be able to see it."

The passenger who did not doubt that we ought to be able to see P., the city he was born in, see it even though the windows were covered with stratus clouds, asked, "What's that down there?" "Where?" we asked. "That gray mass. What desert are we flying over?" "Those are cloud banks," we said. Meanwhile the radio compass needle was receiving signals from P.; we pointed to our feet and said: "P. is down there." The man became desperate. And he said things which made the difference between our vocations glaringly obvious. He described the world-famous joie de vivre of his city. And we flew over it, less than three miles above it, separated from the noisy traffic and bright neon advertisements, surrounded by clouds, clouds and nothing but the clouds for the past seven hours, constantly in their midst, while somewhere underneath us bridges, railroad stations, bars, and art galleries glided away, while we stretched our weary limbs. In this thick, colorless mist, the airplane seemed to stick as motionless as a fly in porridge; the instruments were accurate, our speed was accurate, the variometer showed that we were descending; but we couldn't feel the forward motion, we had no relation to the earth.

And then came the marvelous moment in which the gray began to dissipate; and suddenly, the overpowering blue of the sky burst through. The sky arched in an incomparable distance, mounting from underground fields into azure endlessness. The sky spread out before us as if it had just been created and human eyes had seen it for the first time. Light dazzled our eyes, light that plays on the clouds, pouring out its glow over them until the horizon.

Rudolf Braunburg

It might seem that taking photographs from an airplane is easy. The distance is "infinite," the shutter can remain open, and the light is consequently brief. No great preparations are necessary for oblique pictures since every airplane has a window. For vertical shots, however, such as are needed for cartographic purposes, it is necessary to organize difficult technical requirements. The semi-automatic camera is hung free of vibrations on the outside of the airplane. This camera weighs a good two hundred pounds and constitutes a valuable piece of machinery. The same camera in a satellite produces better pictures because in space there are no vibrations to spoil the focus.

The size of the negative in commercial vertical shots is usually 23 x 23 cm., that of the focus 15 cm. (wide angle with 90° opening angle) or 9 cm. (very wide angle with 120°). The rolls of film are 200 feet long and contain 250 shots. Focal width and altitude determine the scale. It is between 1: 2,000 and 1: 100,000. Usually the scale is selected in such a way that after enlarging the picture two or three times the desired map is reached. Thus, for a map with a scale of 1: 10,000 the camera scale would be 1: 30,000.

Vertical shots for maps call for huge instruments costing as much as a two-family house, and which are basically gigantic stereo-microscopes. Two successive aerial shots overlapping halfway are placed separately in the left and the right ocular. The surveying engineer thereby receives a spatial impression of the area. The enlargement and the stereo view permit him to grasp precisely the location and height of every detail crucial to making the map.

Aerial photography has revolutionized land surveying. Only a few points of reference are measured on the ground; everything else is done with the stereo equipment. Since field work, especially in some almost inaccessible areas, is time-consuming and expensive, we can see the great advantage of the aerial method: the area is brought to the desk as a palpable, three dimensional model. The margin of error is as low as .02 mm.

The complicated surface structures of all industrial countries have been photographically recorded and the rapid progress of building and industrialization can be shown impressively by means of comparisons with older photographs. The aerial photograph is a first-class source of information for other purposes too. For example, in some parts of Africa a census is taken in an approximate fashion by counting the number of huts from the air. And the investigation of unexplored jungle forest areas is carried out by means of aerial photography; a sort of differential diagnosis of these territories can be set up, and we can even estimate the amount of wood that could be gained.

The geological use of aerial shots provides an insight into the construction of the crust of the earth and thereby the possibility of prospecting more directly for petroleum, metal ore, or water. Topographical and agricultural studies can be made more economically at a work-table than in the field; a specialist views the photograph of the area under investigation, and, what's more, in three dimensions. The transforming process from aerial shot to topographical map is one of selection and reduction. The map should contain only essentials for orientation and overview.

The fascinating illustrations in this book are in no wise map-like or abstract. They contain the details that are usually nonessential for the map makers: the joints of a glacier, the bizarre furrows of a field, details that an abstract map cannot show. A cartographer would find the photographs in this volume far too pictorial because they contain too many unessential elements. But for the nonspecialist they are stunning in their novelty and strangeness. They offer something nonpictorial, something abstract in the sense of the modern trend in the arts.

Klaus Völger

The Origins of the Earth

We can only feel space,
when we tear ourselves loose from the earth,
when our support vanishes....
Kasimir Malevitch (1924)

Paul Klee's statement about the modern artist holds figuratively for the observer looking down upon the earth: "With a penetrating vision, he sees things that nature shows him after they are formed. The deeper he looks, the more easily he can stretch aspects of today into yesterday. Instead of a finished image of nature he sees the sole essential image of creation: genesis."

Circular forms, currents, pulsations, clouds of smoke, undulating, semifluid motion: beneath the cool, static surface of the earth—the hidden primal energy. In the rhythm of a still unknown periodicity, cosmic signals light up, the signs of the earth's interior. The crust vibrates, fiery gases stream away. Volcanic islands rise from the bottom of the sea, recalling the most ancient myths which claim that the earth and all life come from the water.

The panorama of creation includes the angular joints of powerful glaciers, the chiaroscuro paths on which the ice masses are sluiced through mountain formations, glacial landscapes, estuaries and deltas of mighty torrents, the thousandfold veins and arteries of river areas, ornaments of the movement of energies and processes on the surface.

The levels, folds, surface motions of the earth's crust belong to its inner dynamics. A stony primal landscape, curvaceous forms, tectonic figurations of rocks, tropics and jungles, steppes and deserts —the history book of the world, lies open, huge and readable, below us in the depths.

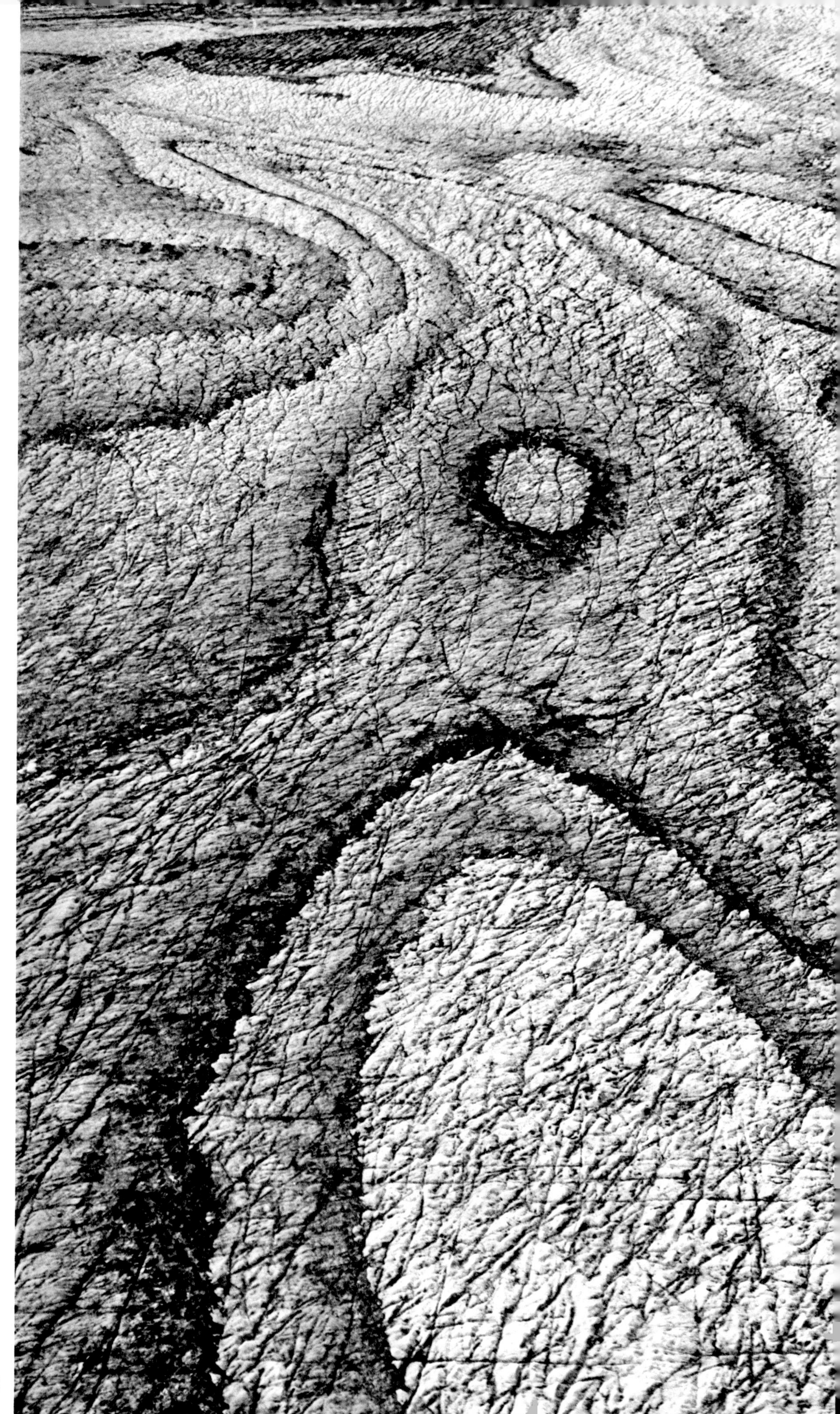

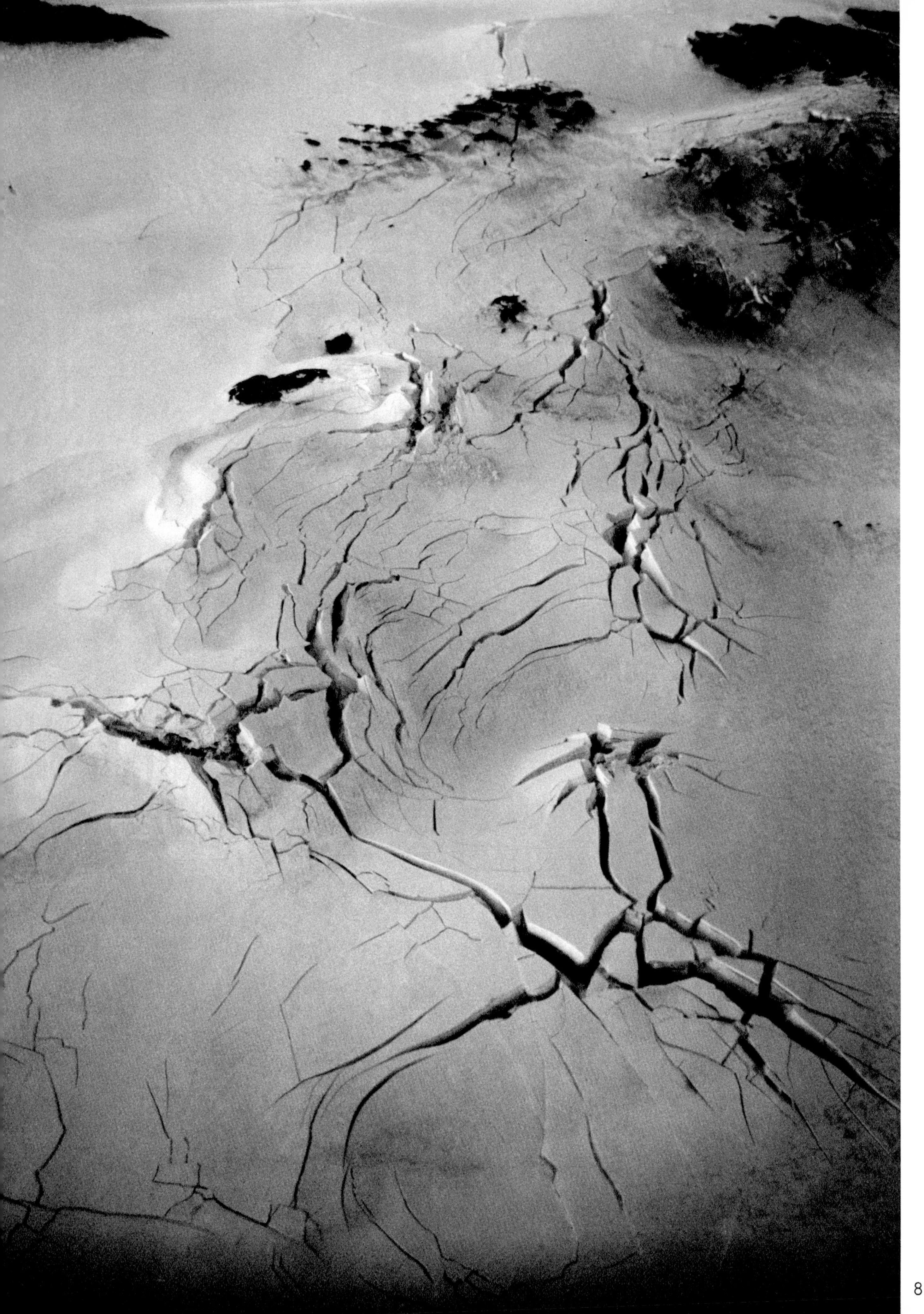

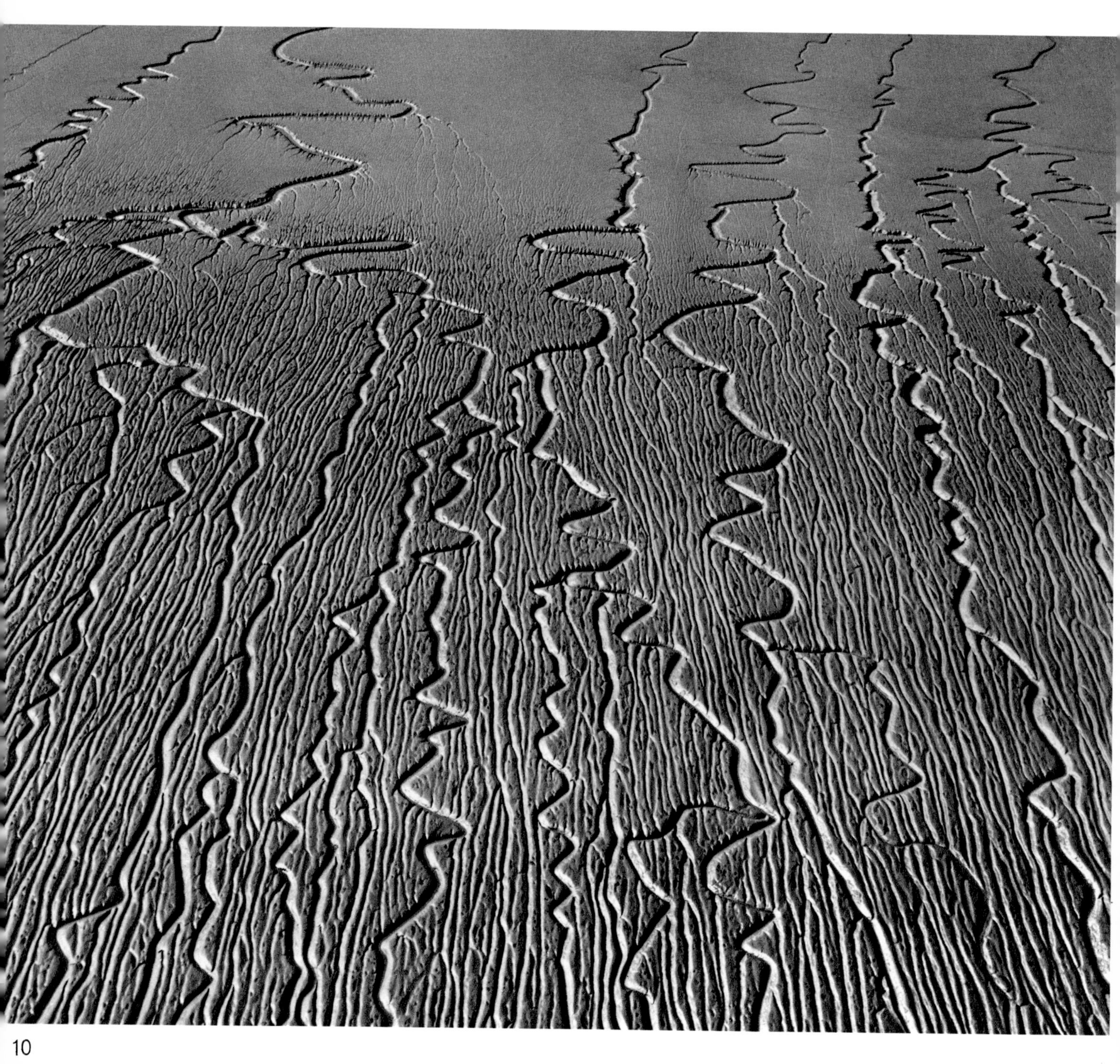

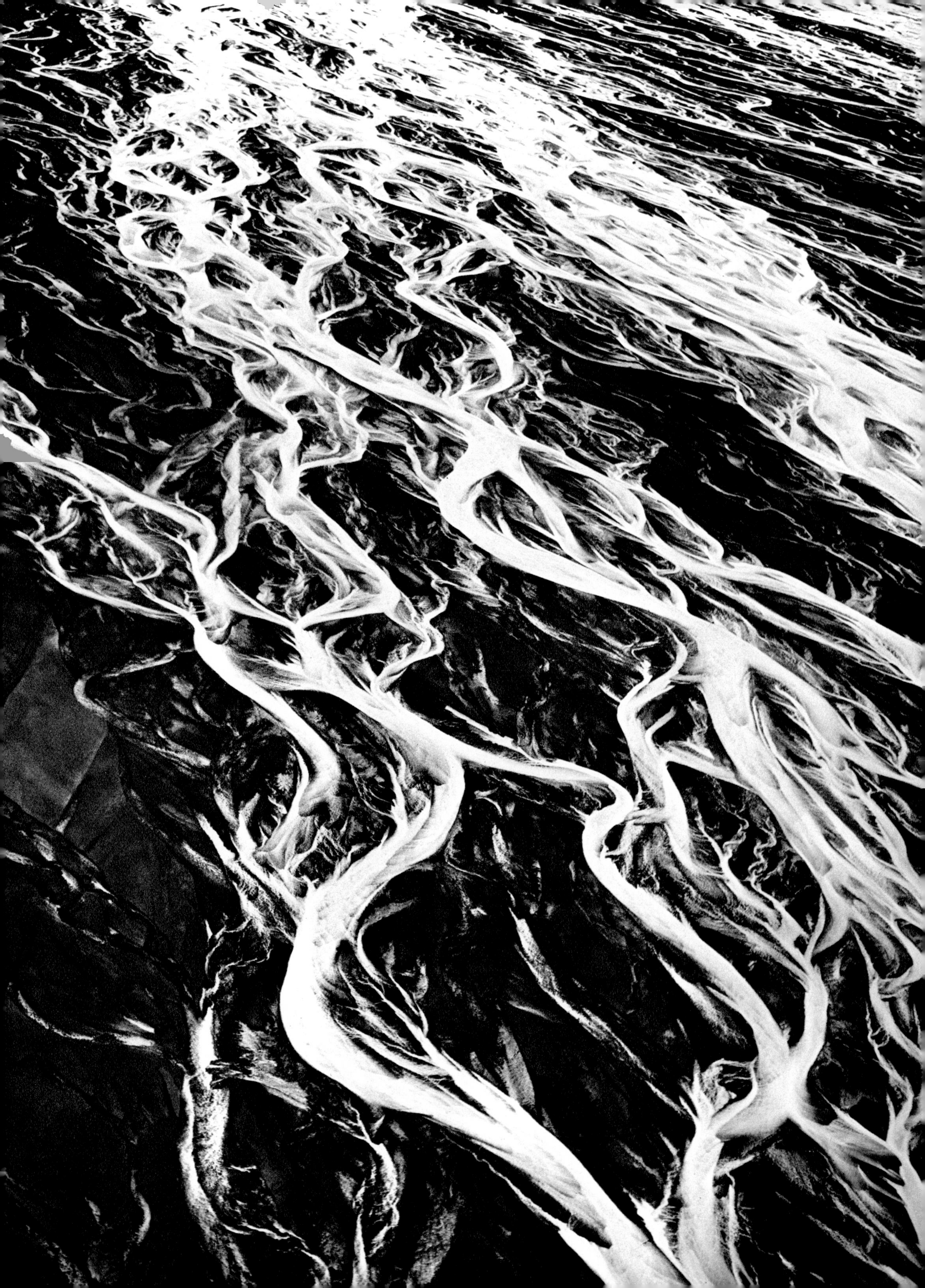

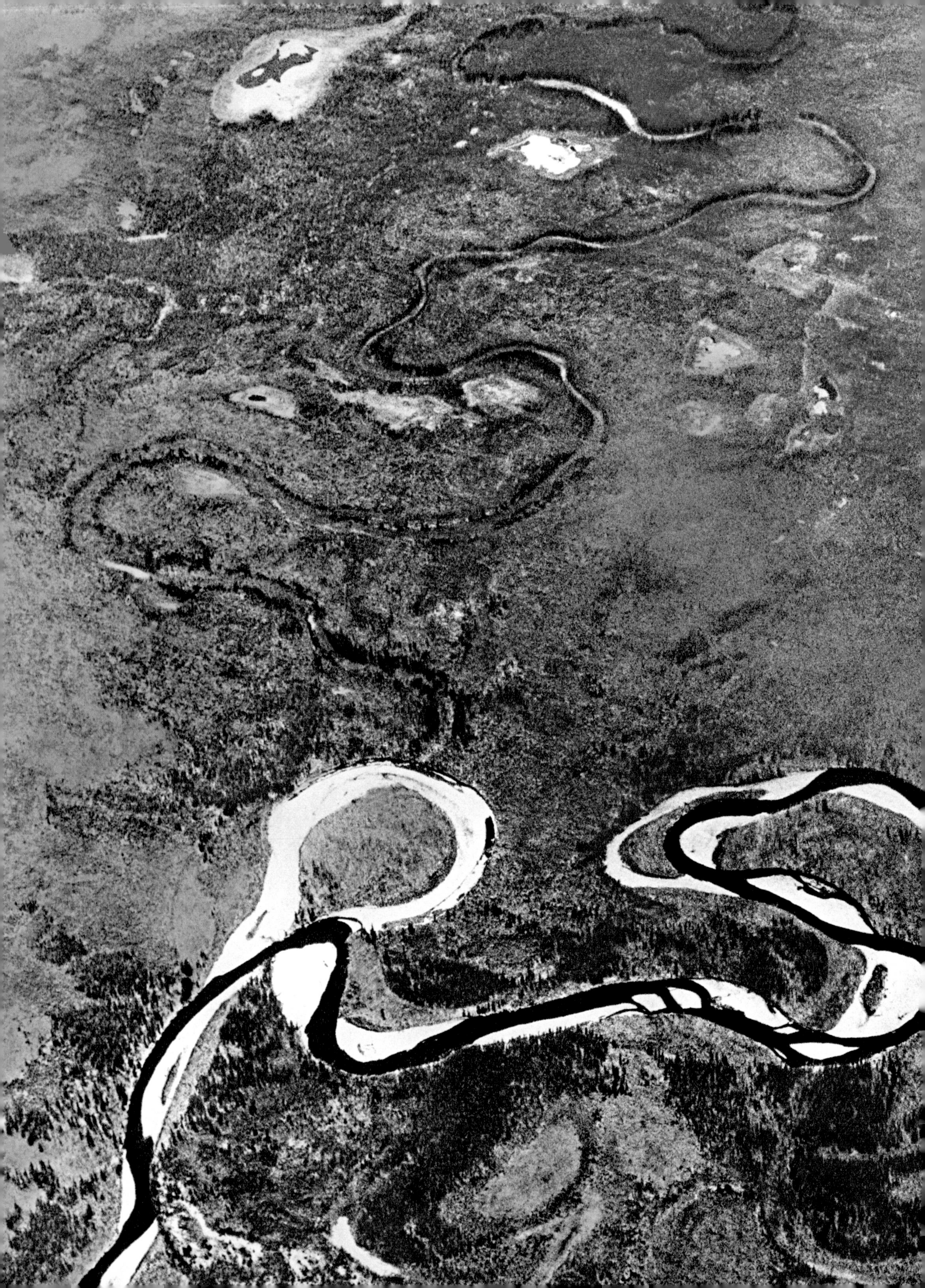

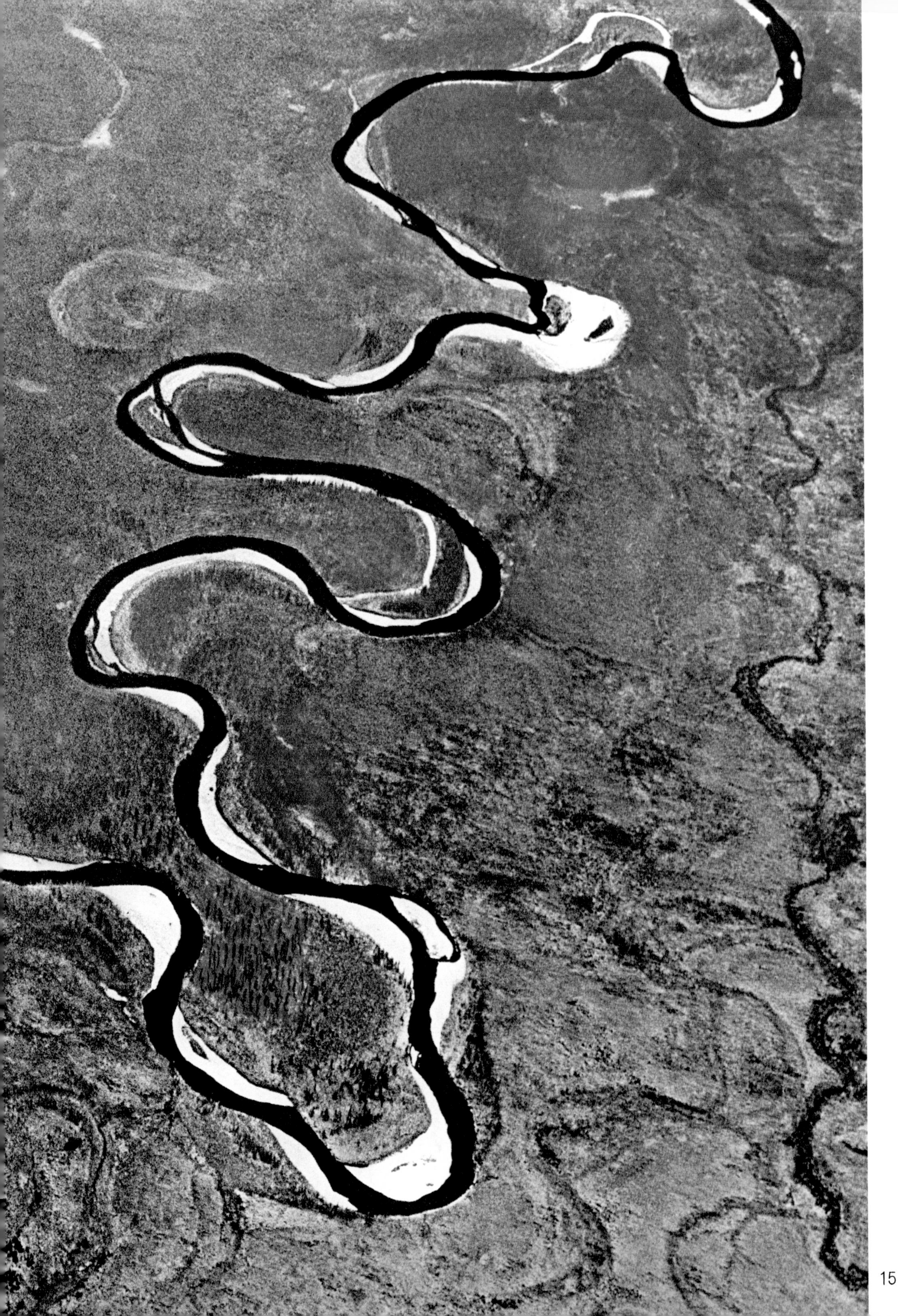

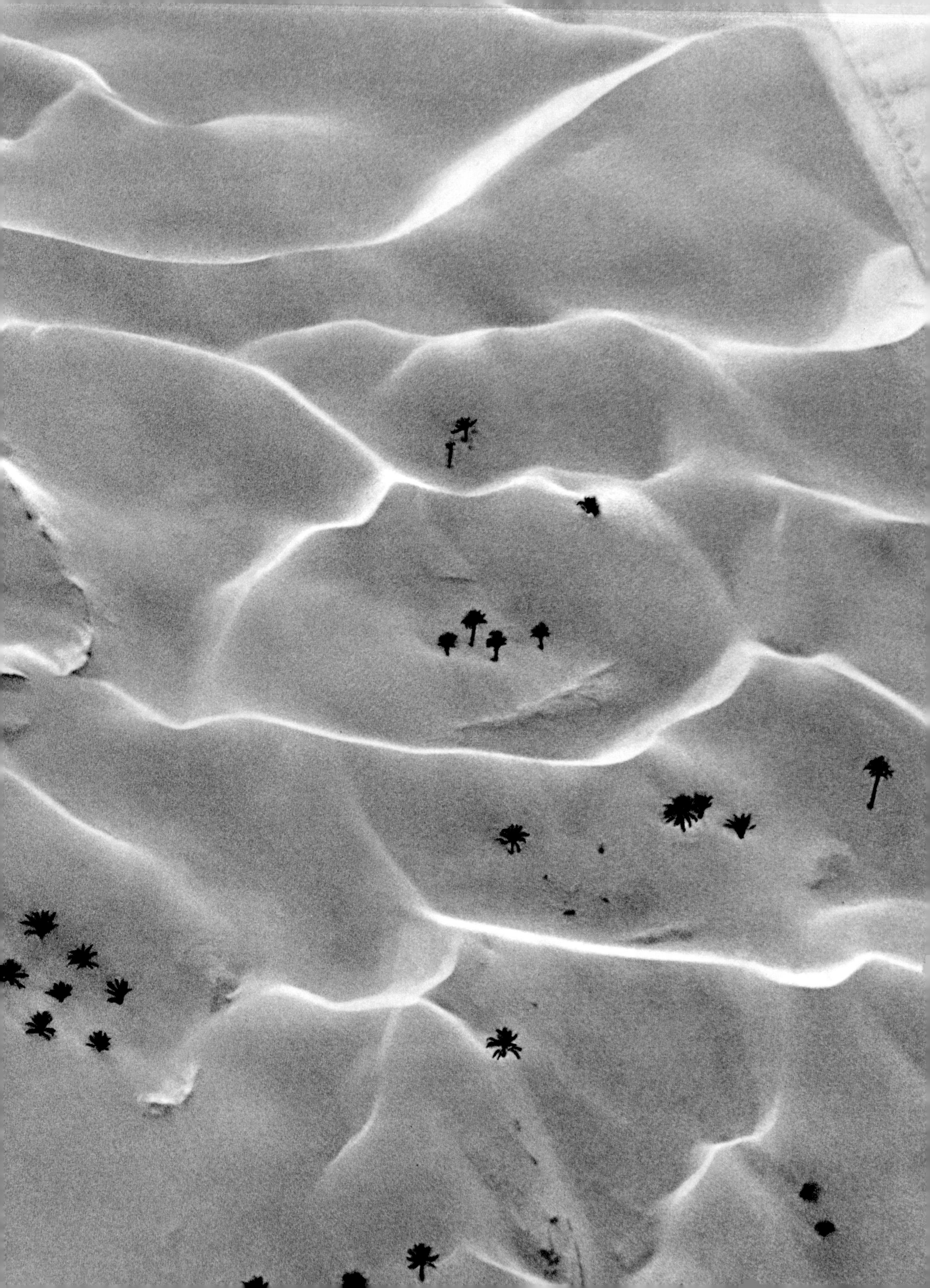

MONTMARTRE
LES TERNES
Eglise Russe
Parc Monceau
Arc de Triomp
Avenue du Bois de Boulogne.

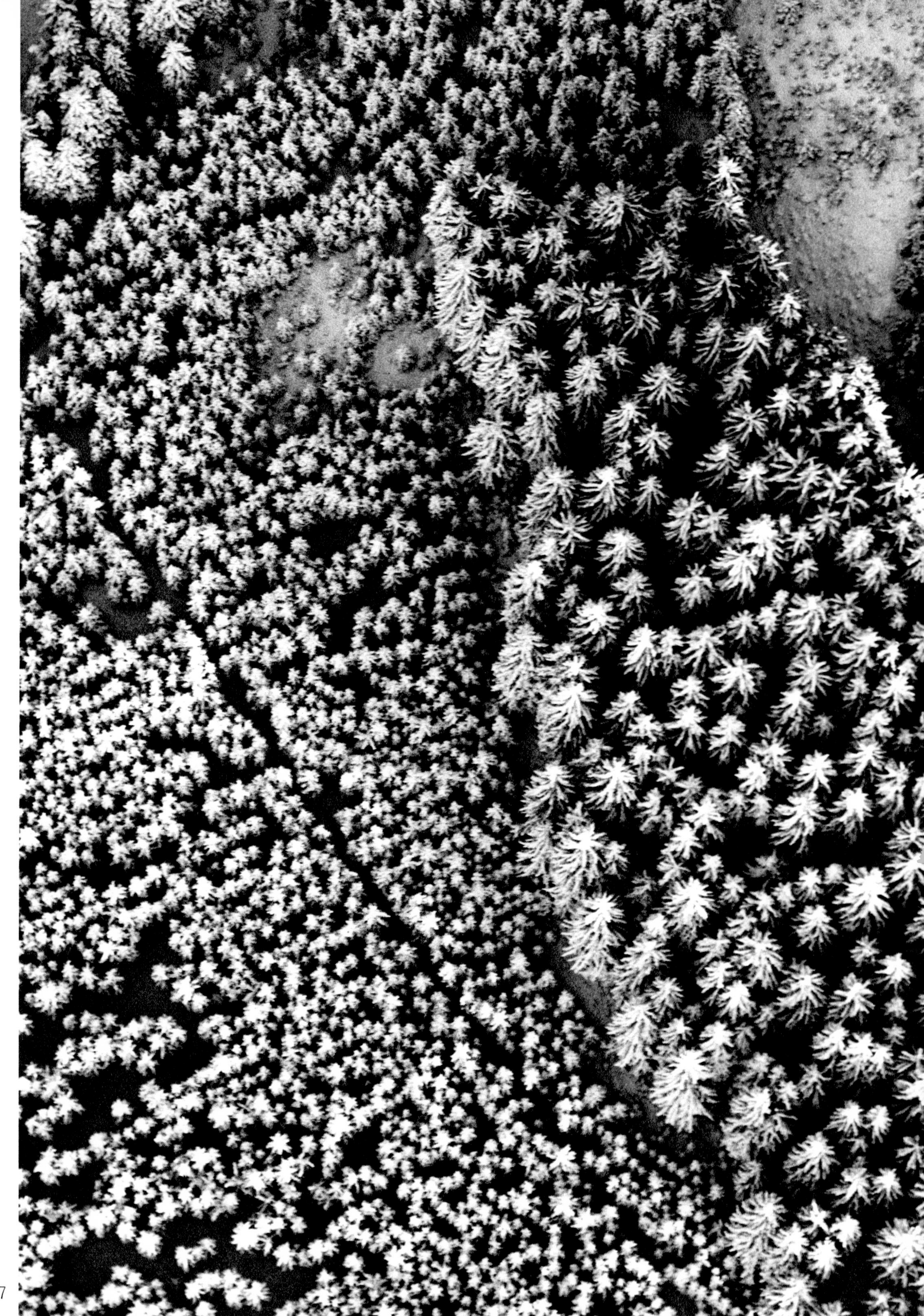

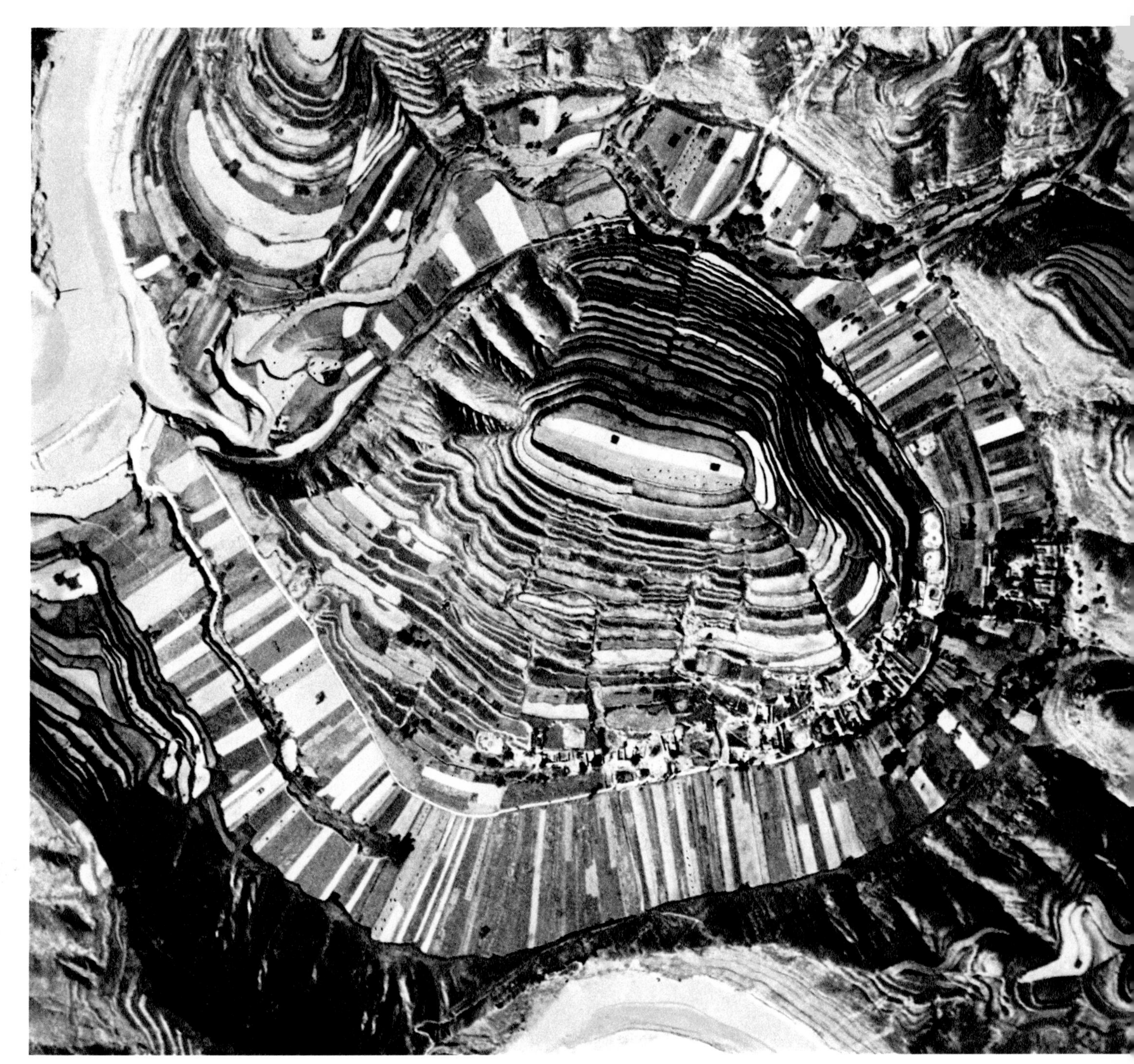

Urbanization and Human Housing

The aerial camera with its x-ray eyes penetrates the skin of the earth and registers vanished life. A modern archaeologist climbs into an airplane in order to deciper the runes of past civilizations. The hue of the soil formations, shadowy signs brought into relief by slanting lights are discerned by the "hyperperiscopic" camera. Circles, squares, rhombuses, and stars are the basic forms of the burial places, temples, army camps, sacred buildings, and other monuments of humanity–three circles: prehistoric tombs; a cubic block covered with crosses: a stone church in Ethiopia; a square embedded in a forest: a Hindu temple; a starry radiance: a Renaissance city; Manhattan: crystalline symmetry of gigantic planes in rhythmic harmony.

Despite the uniqueness of his existence, man is no longer discernable as an individual. In the shrunken world of a deobjectified landscape he can only be seen as a mass. The rapidity of flying makes evident the simultaneity and juxtaposition of civilizations: bent, spiral paths; primitive mountain serpentines; the rhythmic and functional power of many-storied highways and bridges in modern mass traffic. Deserts, jungles, mountains, and polar areas have been viewed and recorded from airplanes. On the planet earth, there is little terra incognita left.

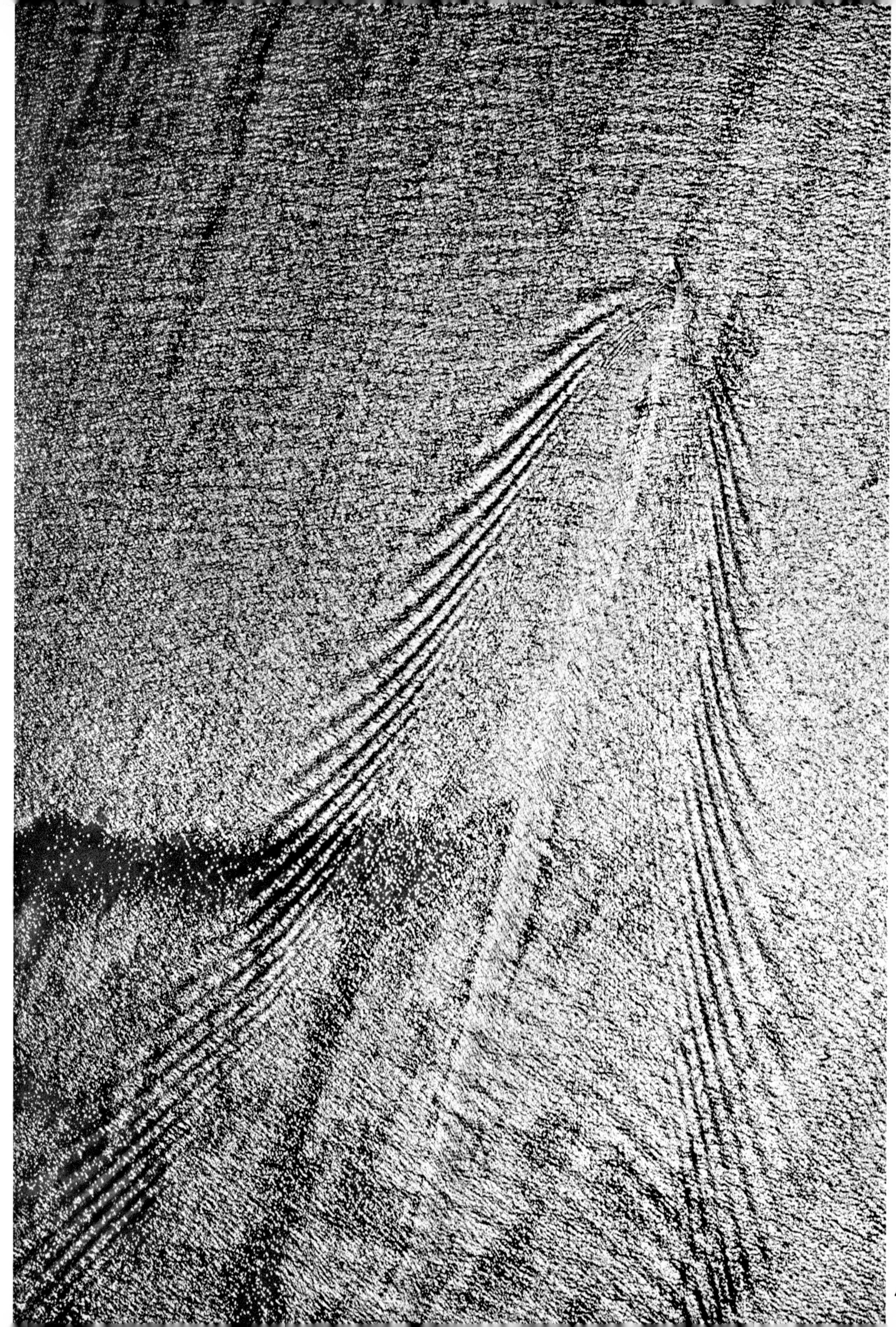

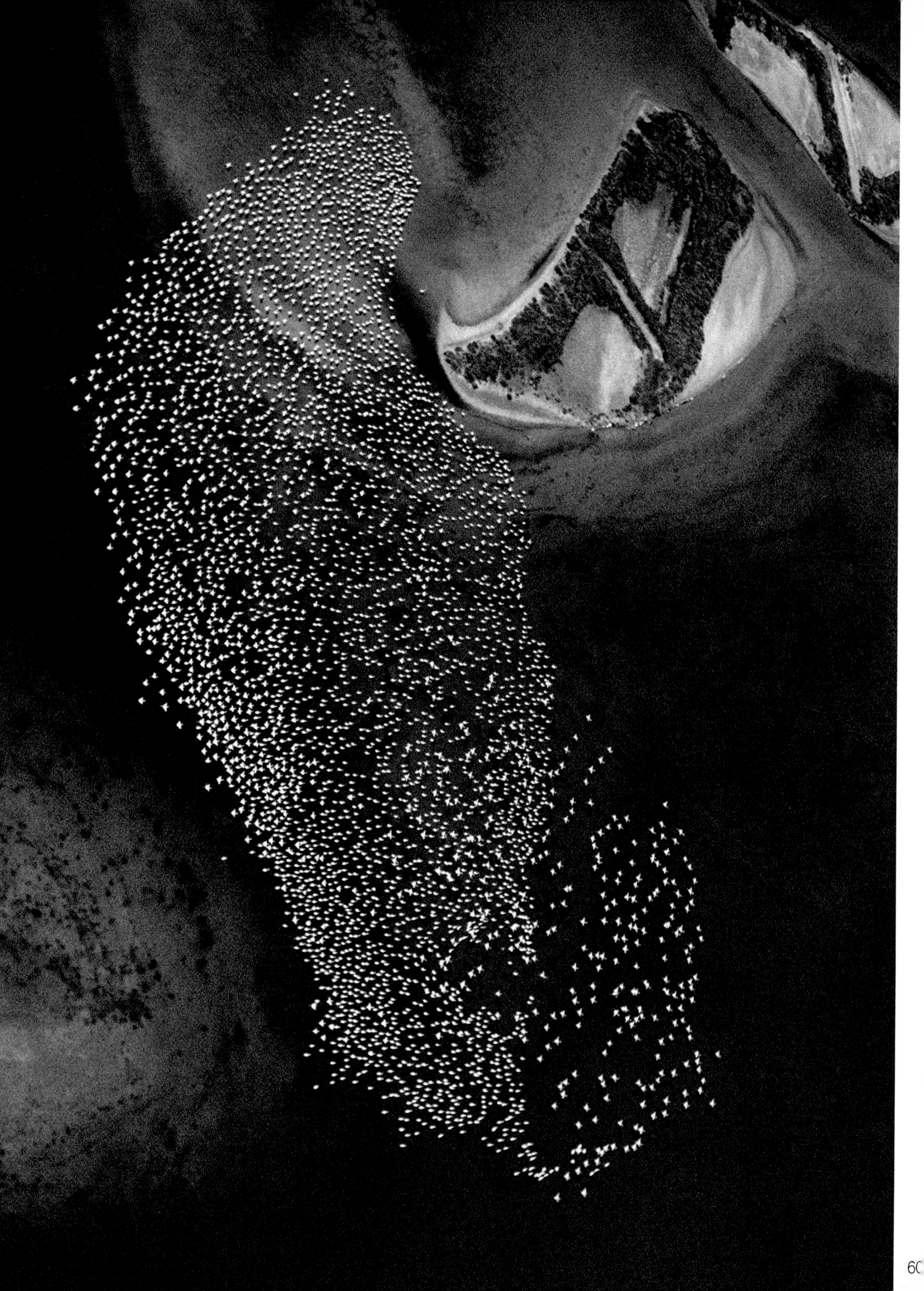

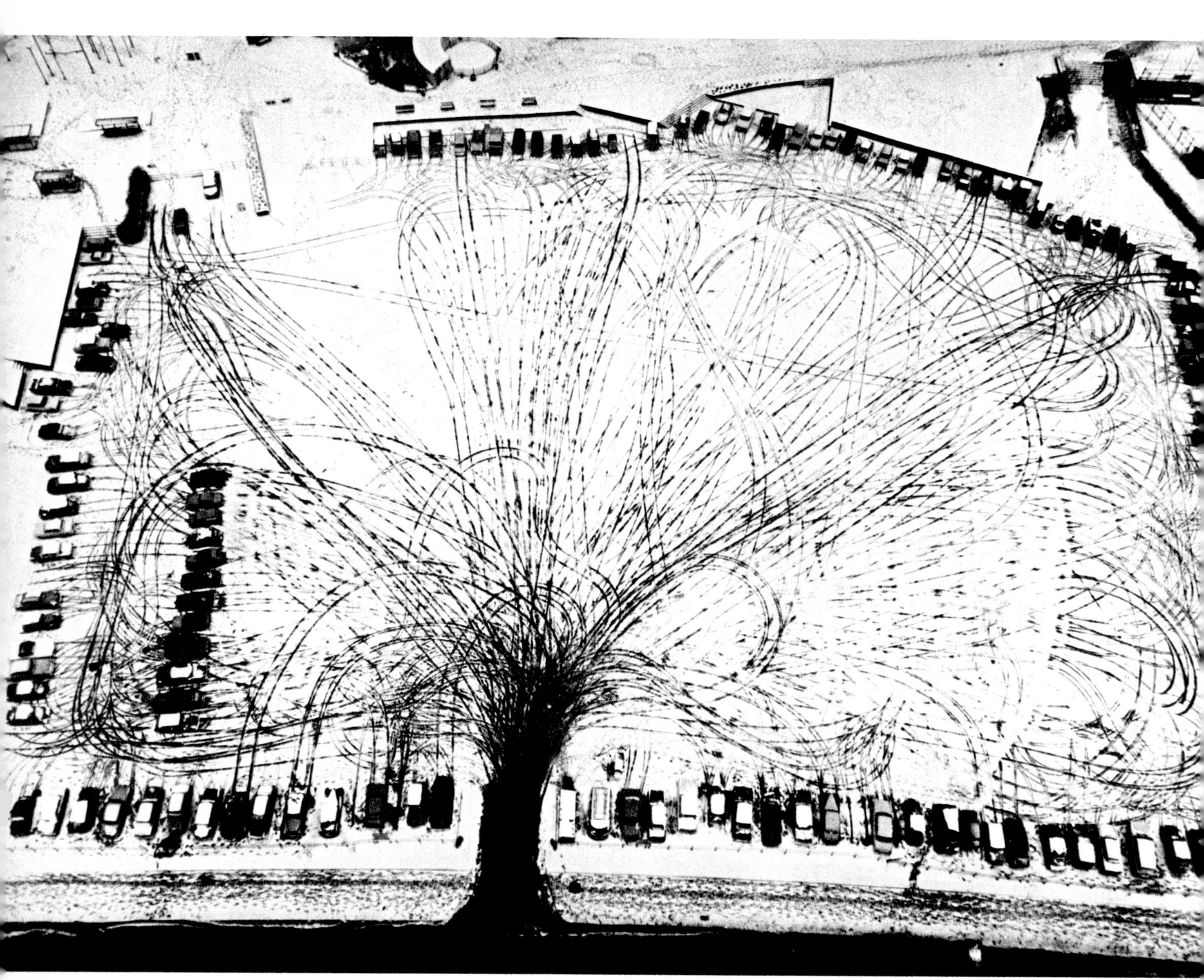

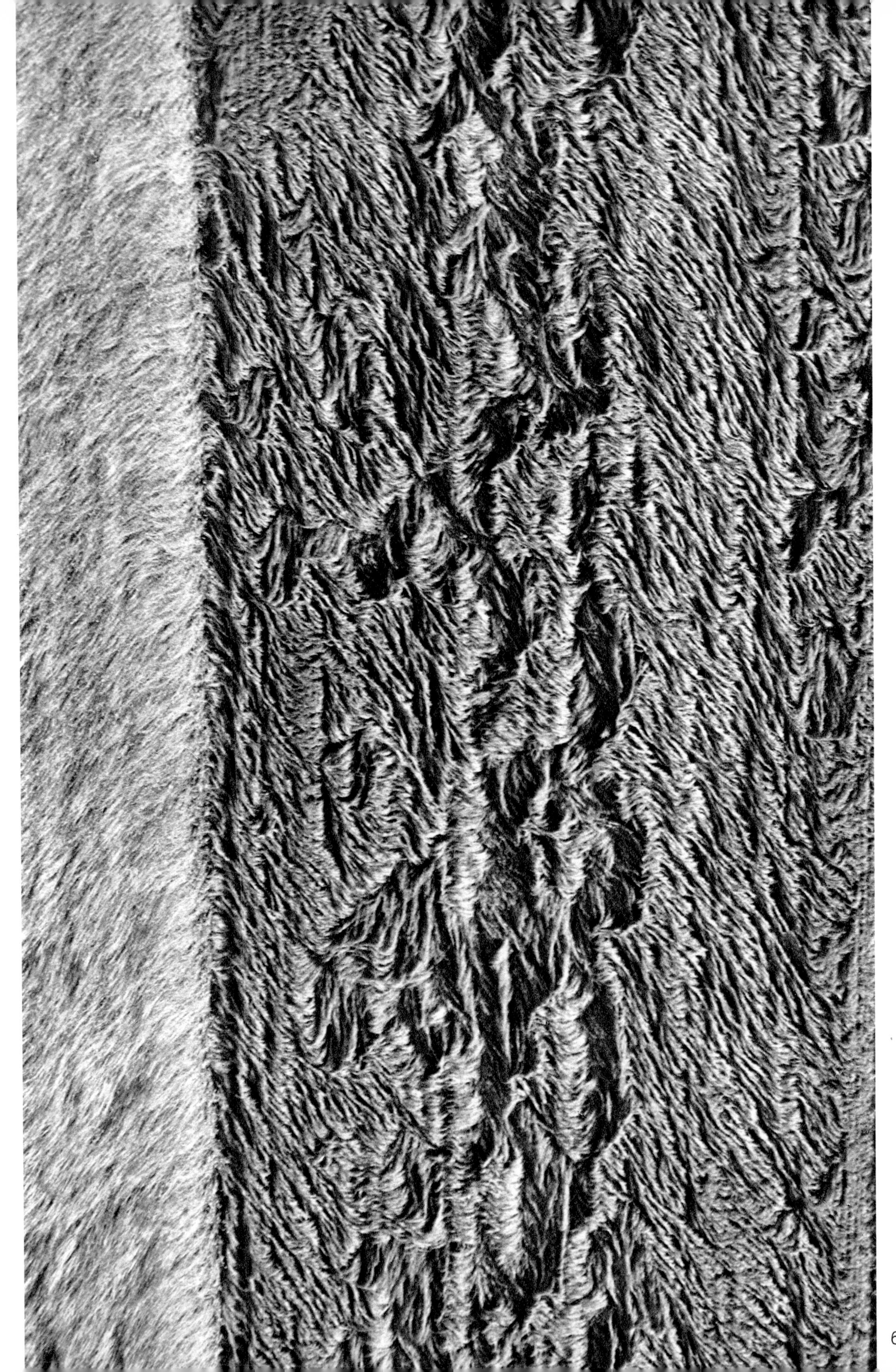

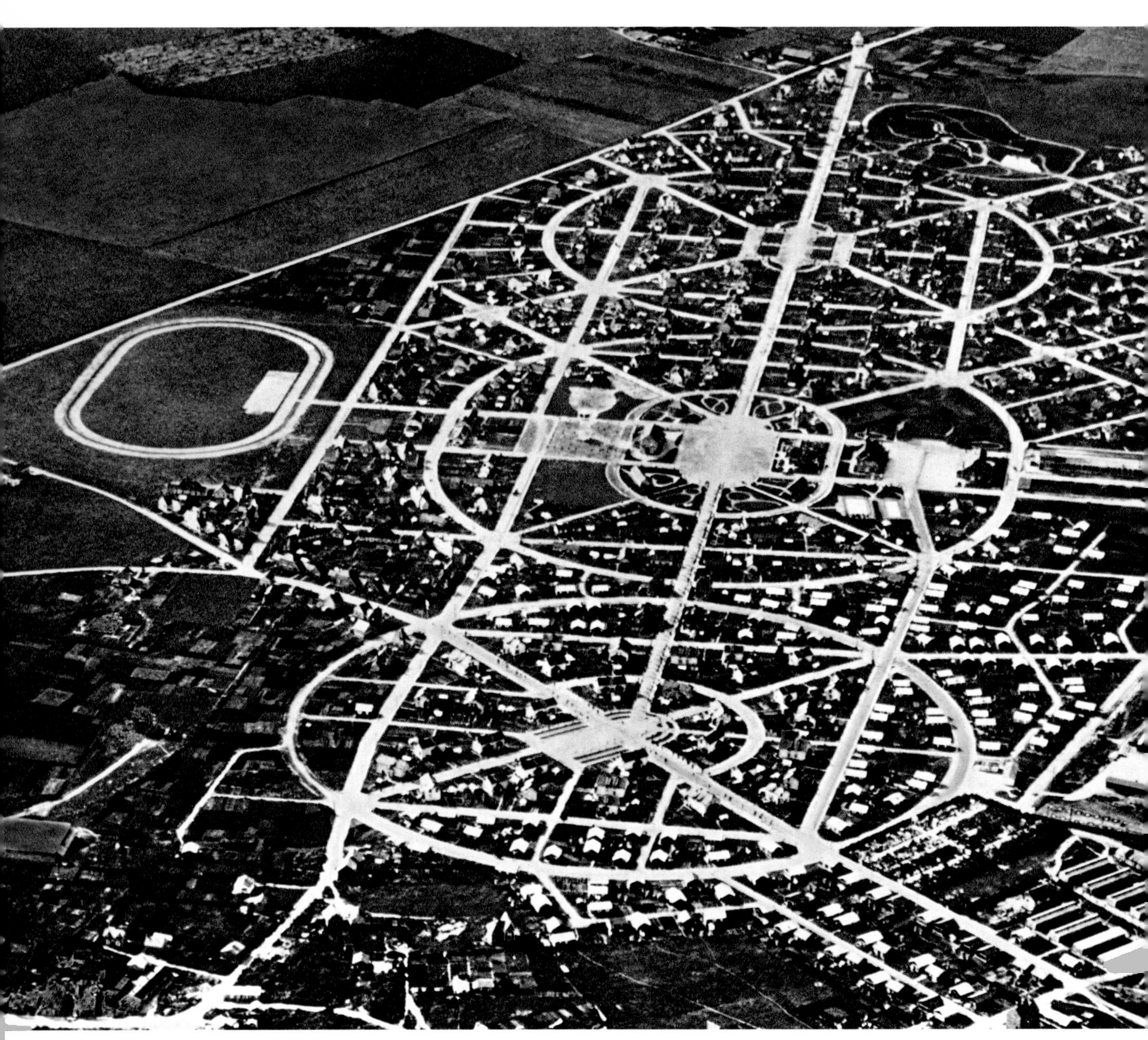

Vision of Art

In the twentieth century, we shall dwell among alien faces, new pictures and undreamt-of sounds.

Franz Marc

By freeing himself from the earth and viewing the world from above, man not only sees more than he has previously, but he sees it in a different way: individual things diminish, vanish, submerge in the general abundance. Land formations, the reliefs of soil areas, the crisscross of waterways, the hues of the earth's carpet, the technical ornamentation of industrial landscapes, and the urbanistic geometry of cities, when seen from the air, recall the paintings of modern abstract and optical artists.

Vertical shots of huge communities resemble the rhythmical arrangements of Piet Mondrian, or of the founders of geometric abstraction. The photos of fields with plowed-up curves that protect the soil from eroding bring to mind the orphic circles of the painter Delaunay. Infrared pictures of territory divided into strips, linear systems and parallels remind us of Paul Klee's paintings of landscapes and gardens. Both in life and in art, the texture of the earth and everything that grows upon it glistens and shimmers.

Seen from aloft, the white ribbons of wood trunks are similar to the linear ditches of optical and kinetic art. Trees forming round rafts near the dark forest and the surface of waters resemble the panes of Kenneth Noland. Like a willow or an olive tree, bowed and twisted, a river flows with its subsidiaries, as if it were painted on silk by a Japanese artist. The twists and turns of labyrinthal arabesques, cut into the soil by plows and mowing machines, evoke associations with Willi Baumeister's paintings of growth. An aerial photograph often resembles paintings or drawings. It is almost as if we were looking at an imaginary museum of abstract art.

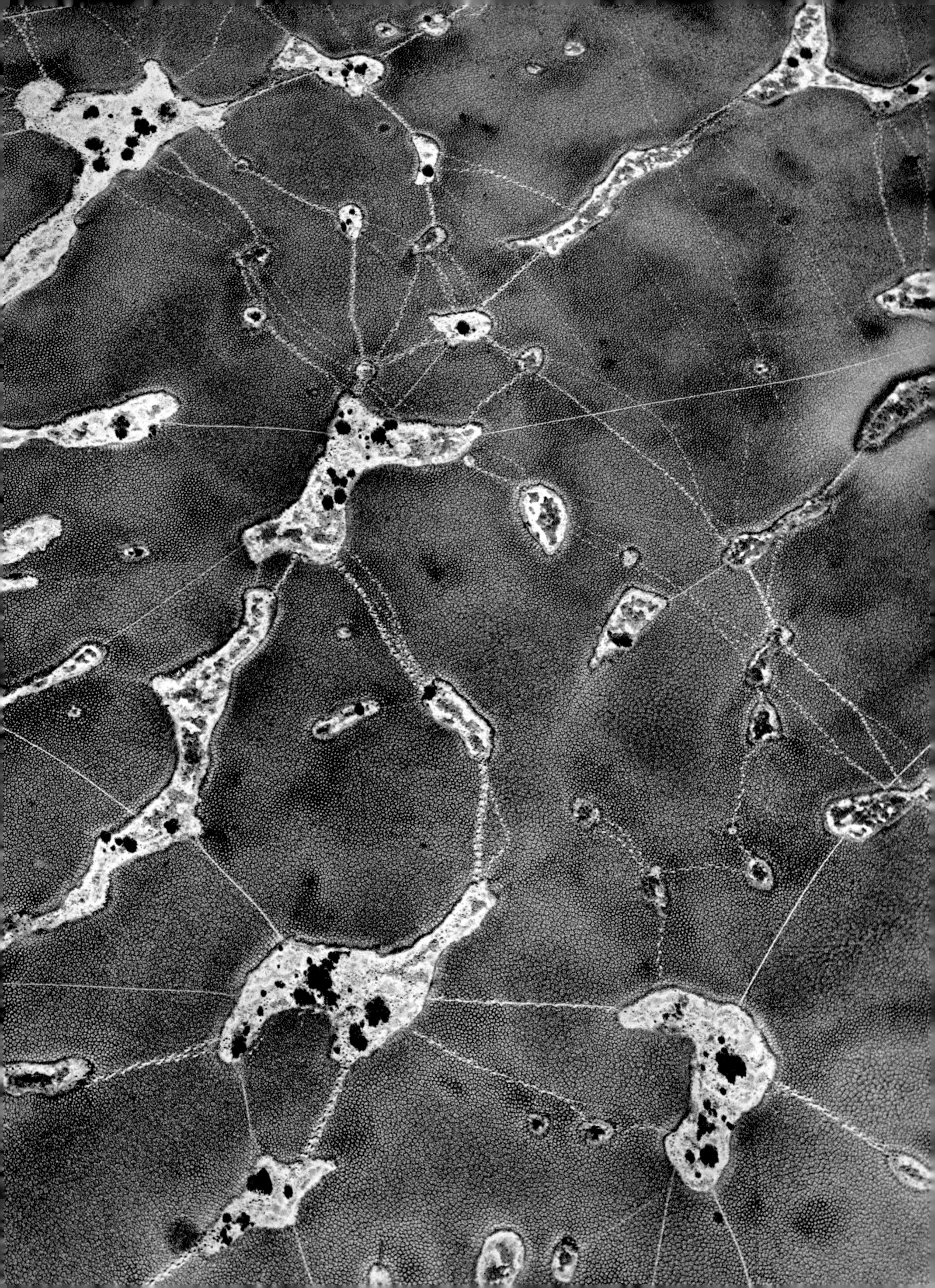

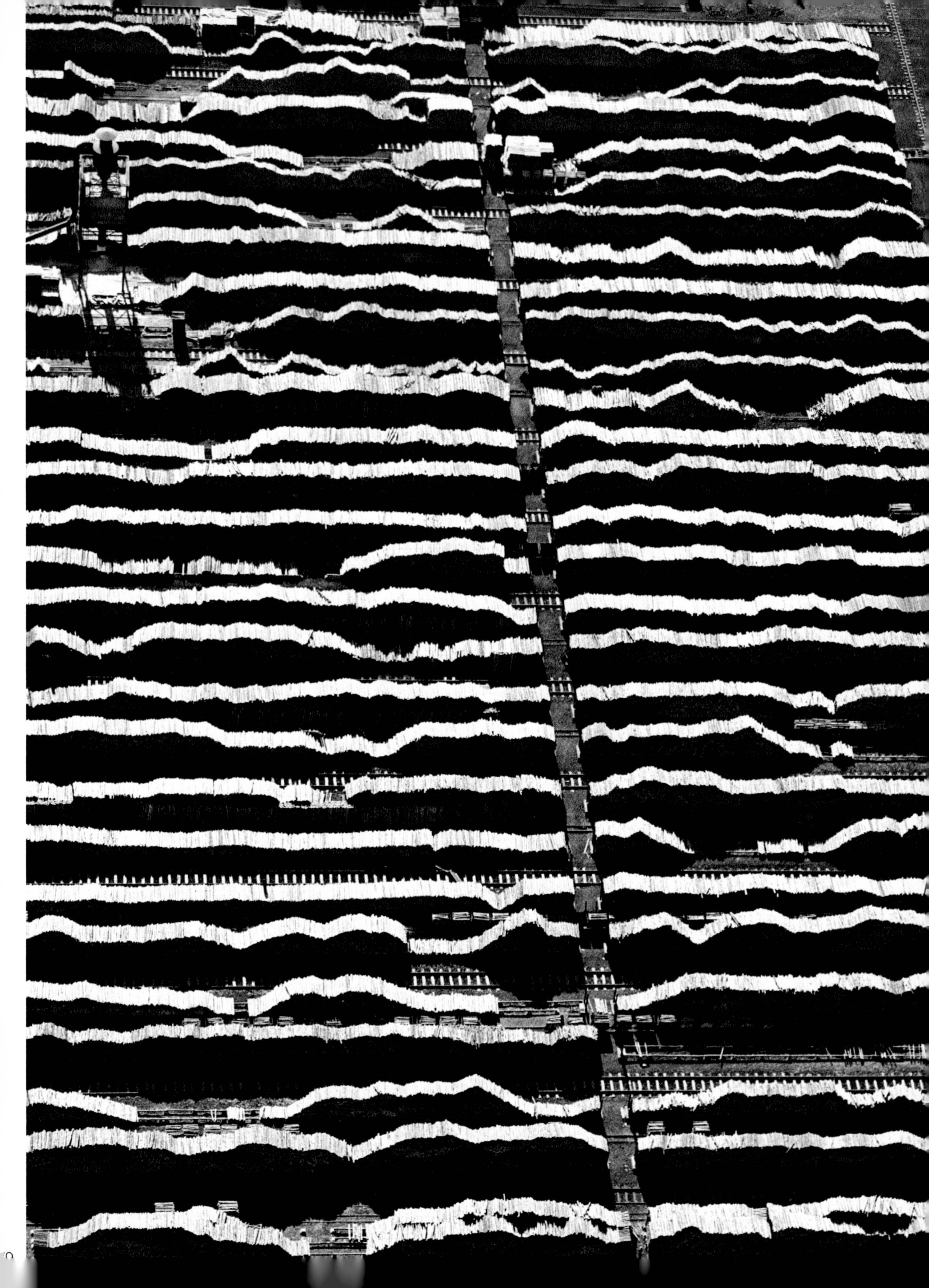

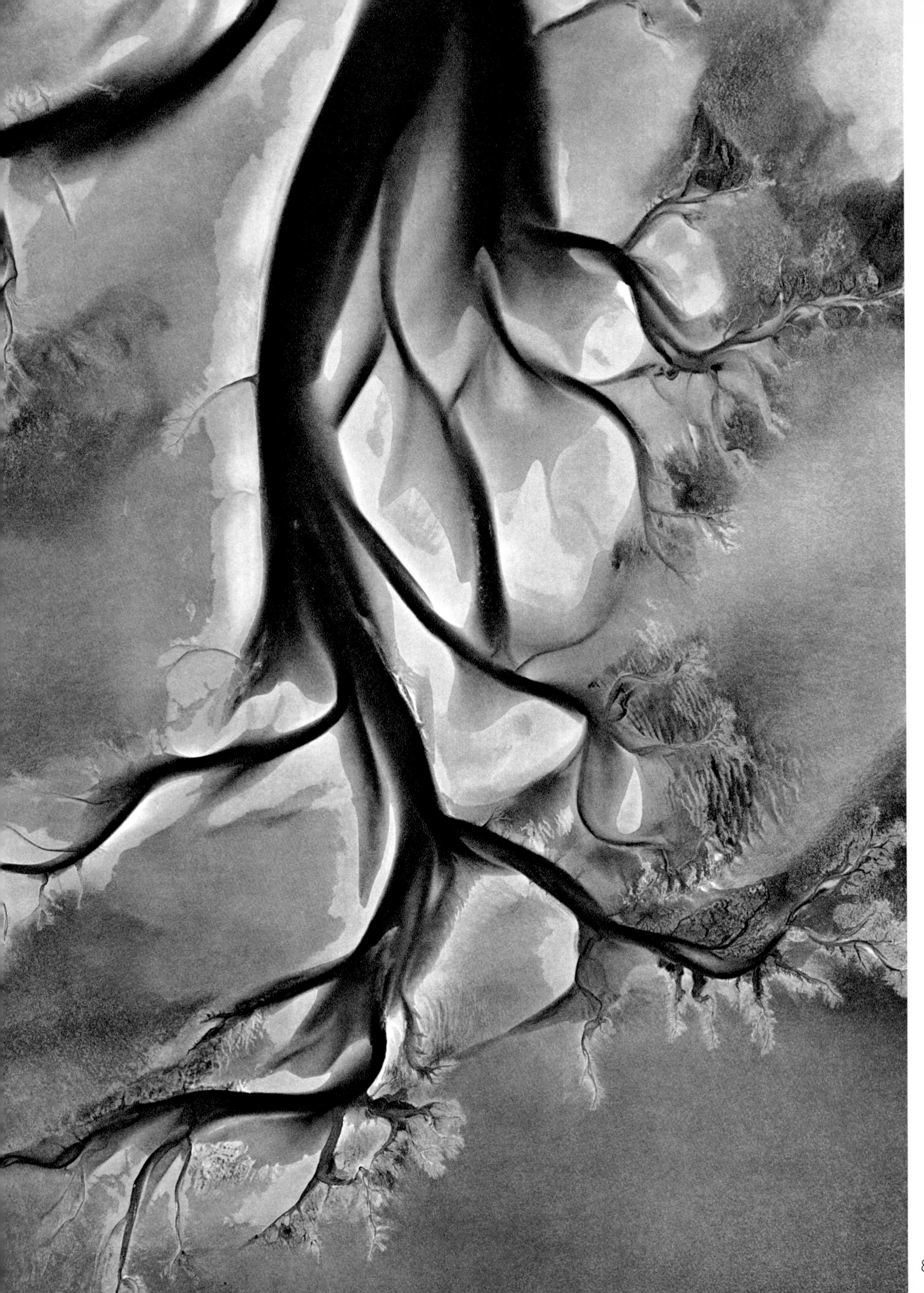

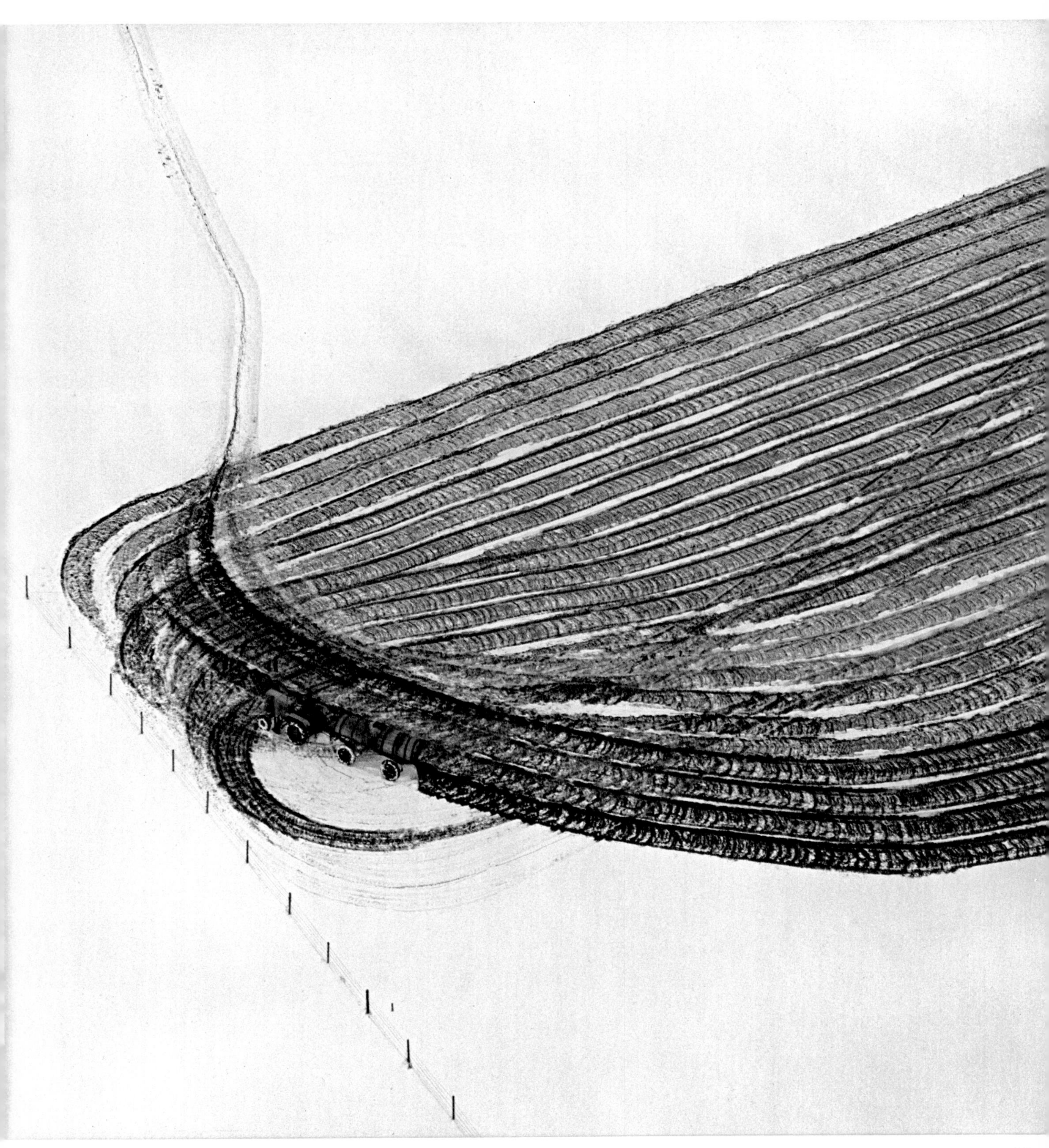

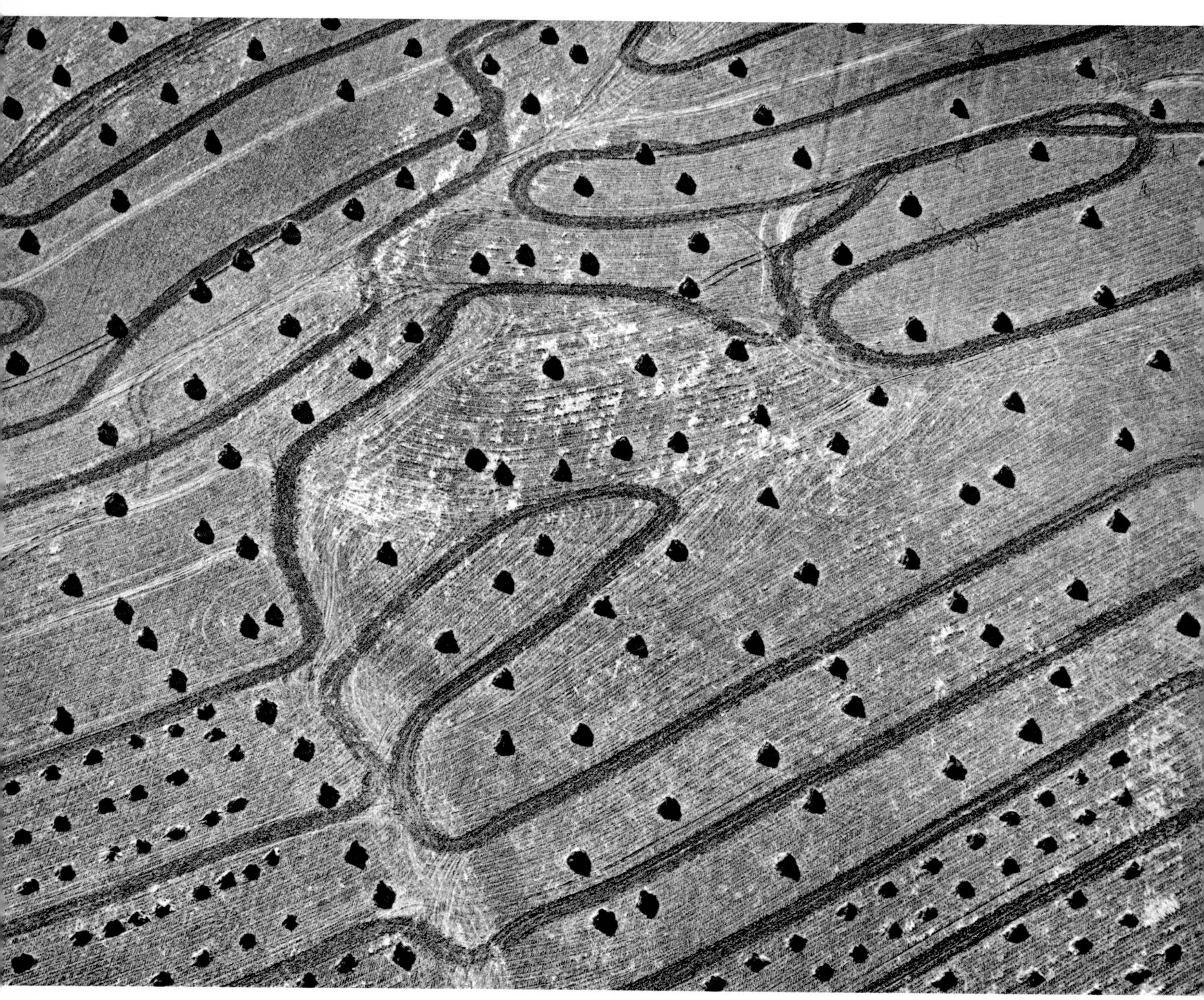

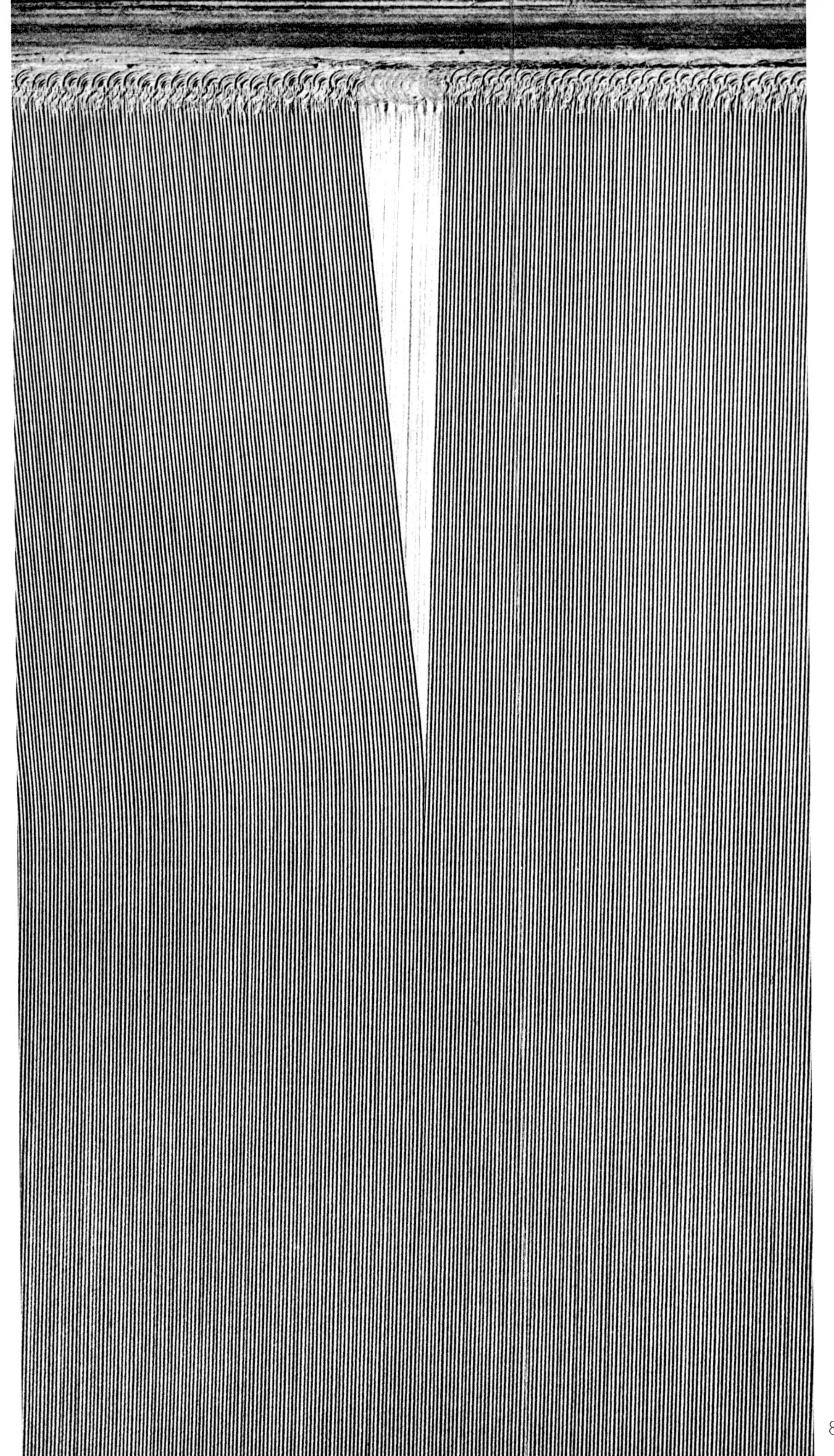

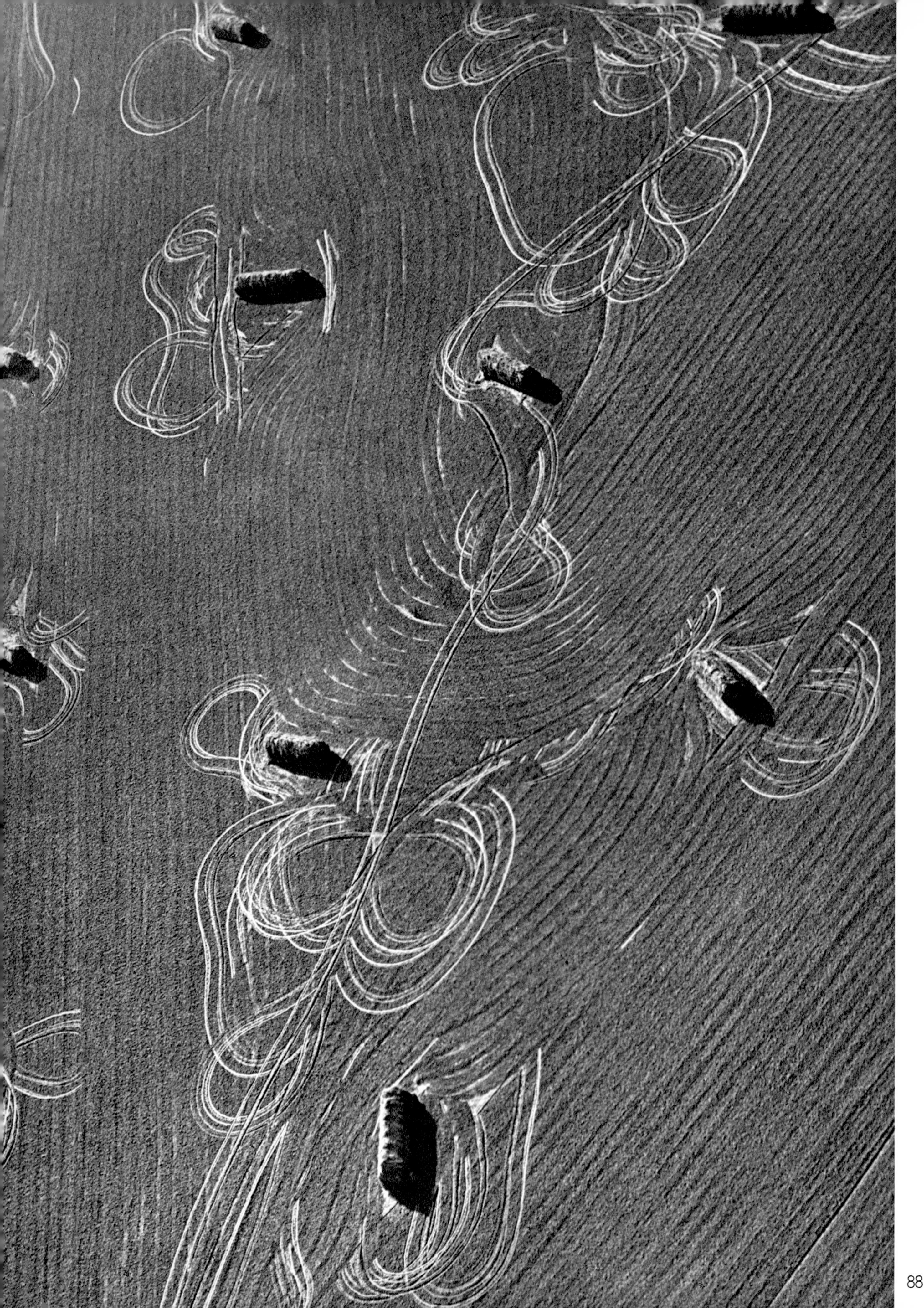

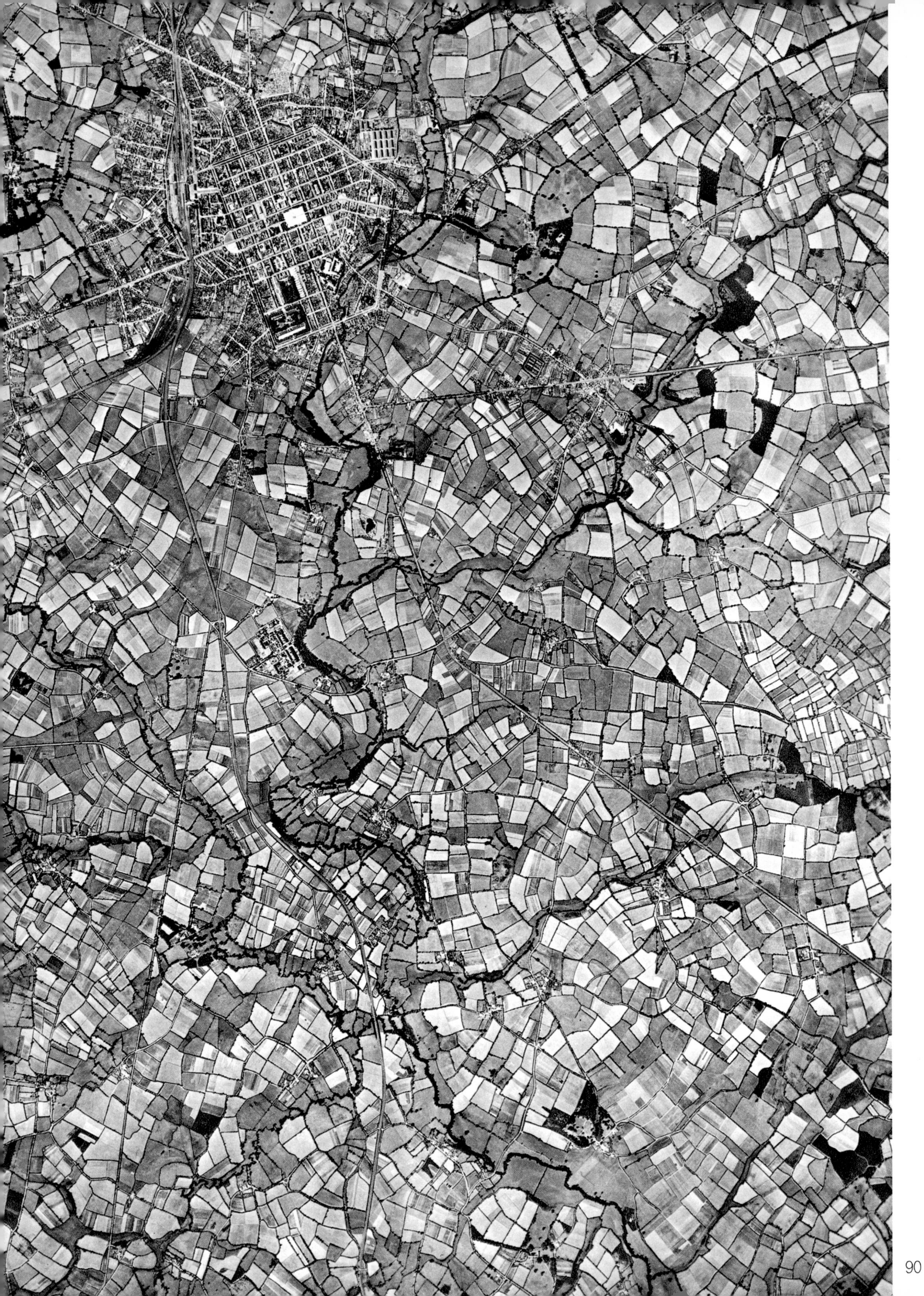

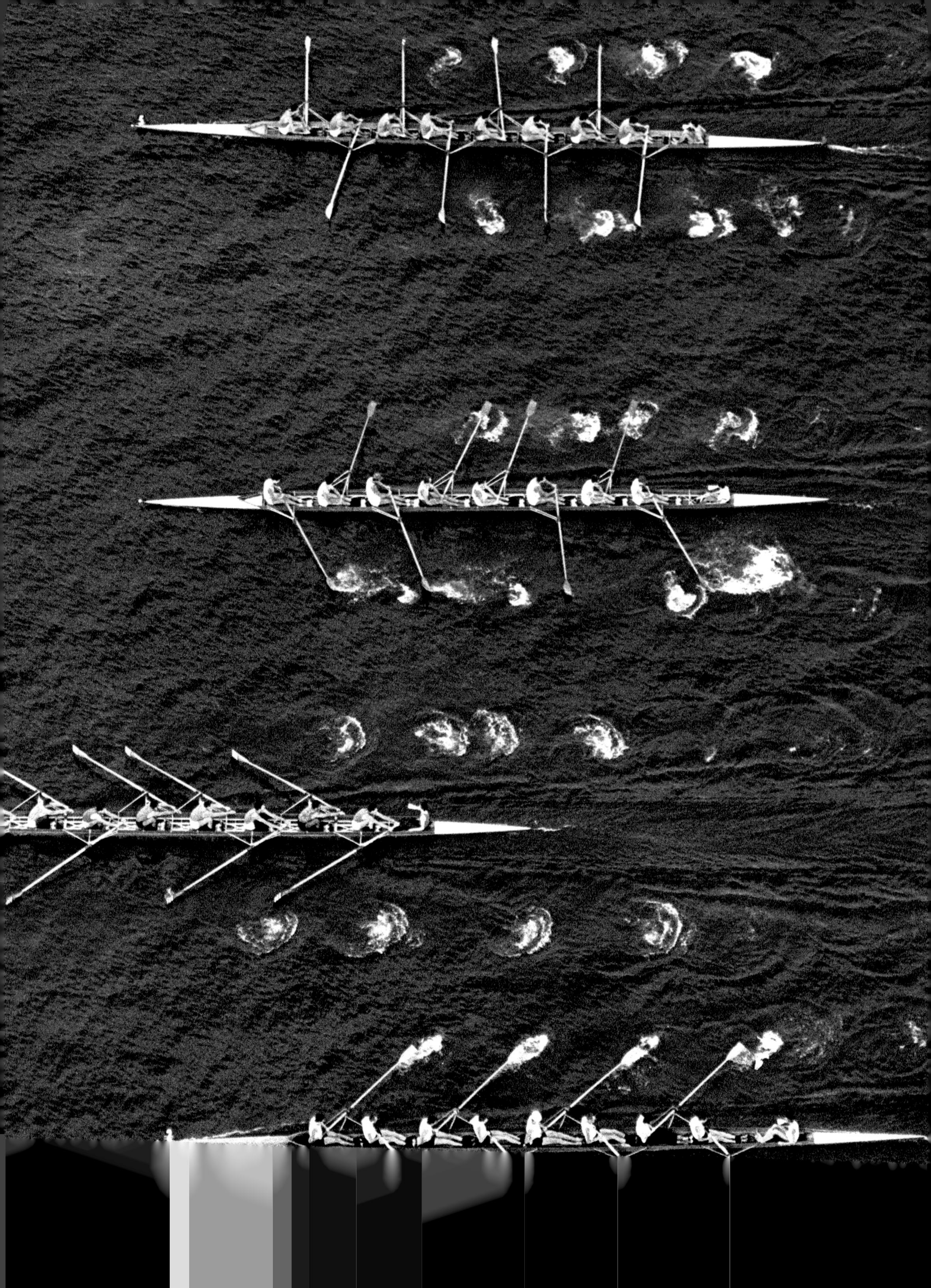

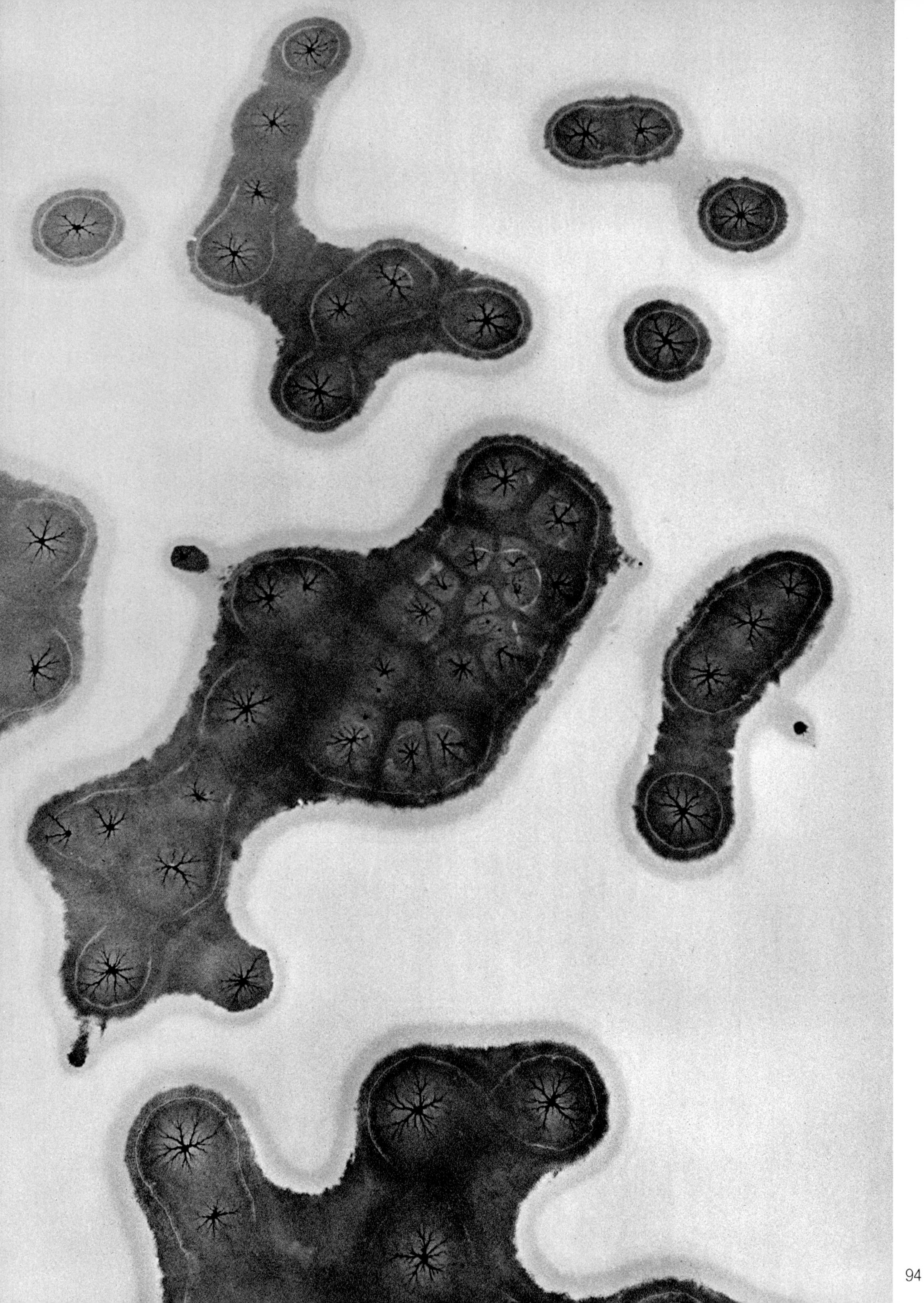

Pictorial Reports from the Universe

We can never discover new continents
unless we have the
courage to lose sight of all coasts.
André Gide

Above the atmospheric roof of the world, satellites revolve. They transmit serial pictures from faraway space, photographs of the sun and the planets, the spiral nebula, the Milky Way, and the dark side of the moon.

Man, who hitherto has moved about only on the surface of the earth, a captive of its gravitational force, can now send out reconnoiterers to circle the globe and to bring back information and pictorial reports from the universe. The myth of Icarus was born of the conception of human flight with artificial wings. All types of aircraft development were based on the model of bird flight. But the satellite is beyond all classical models of flying. It glides about like the earth, like the stars and planets.

Soviet and American space pilots have viewed and photogaphed the cosmic landscape. Spiral nebulae moved from their decorative and poetic distance into man's range of vision. The astronaut contemplates the earth from outside and far away. He recognizes the earth's countenance only in great lines: the circular mirrors of the oceans, the dark masses of wandering mountain chains, the pale dominoes of cities, the bending of the earth's surface, and the dimly glowing moon.

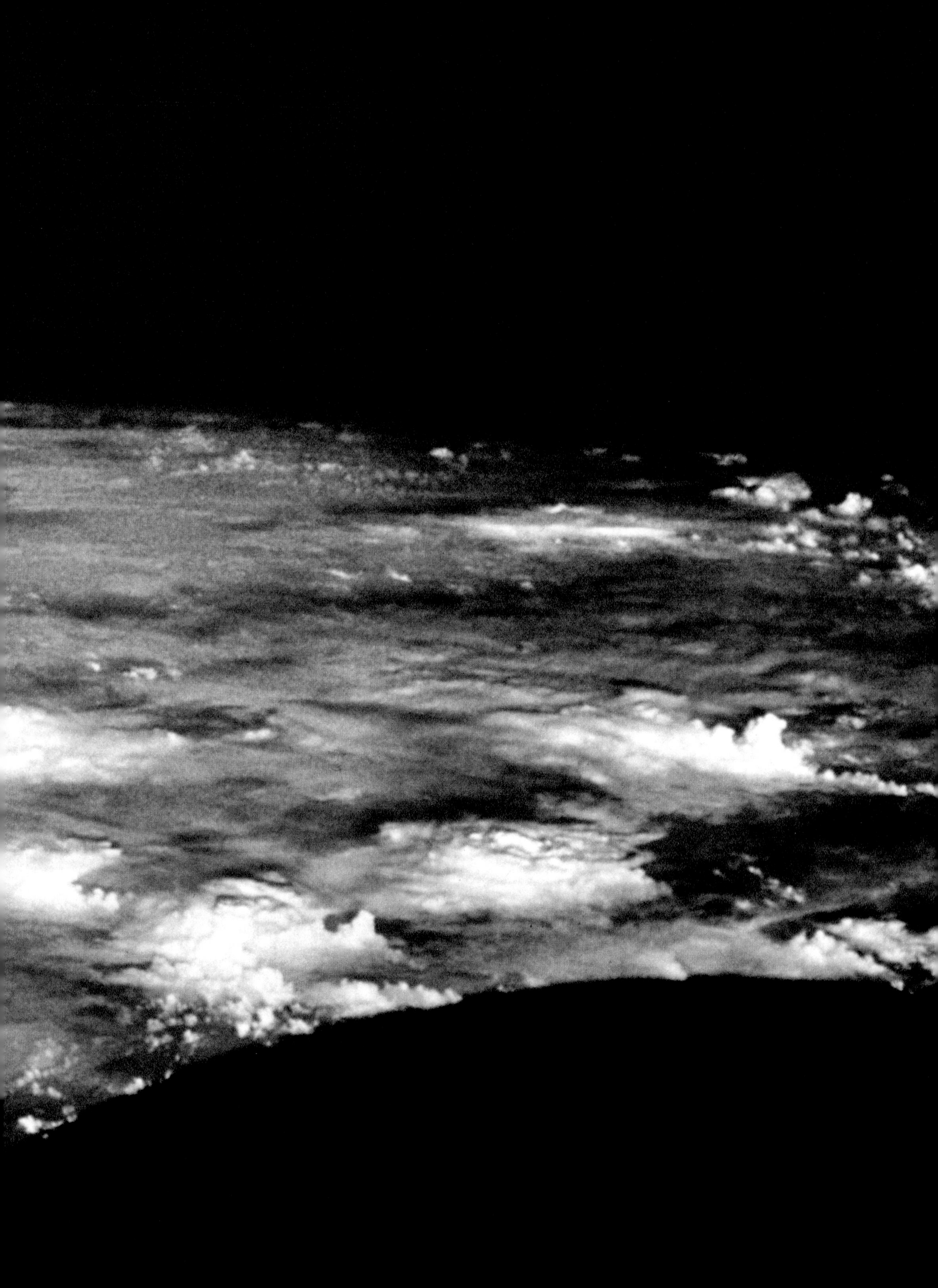

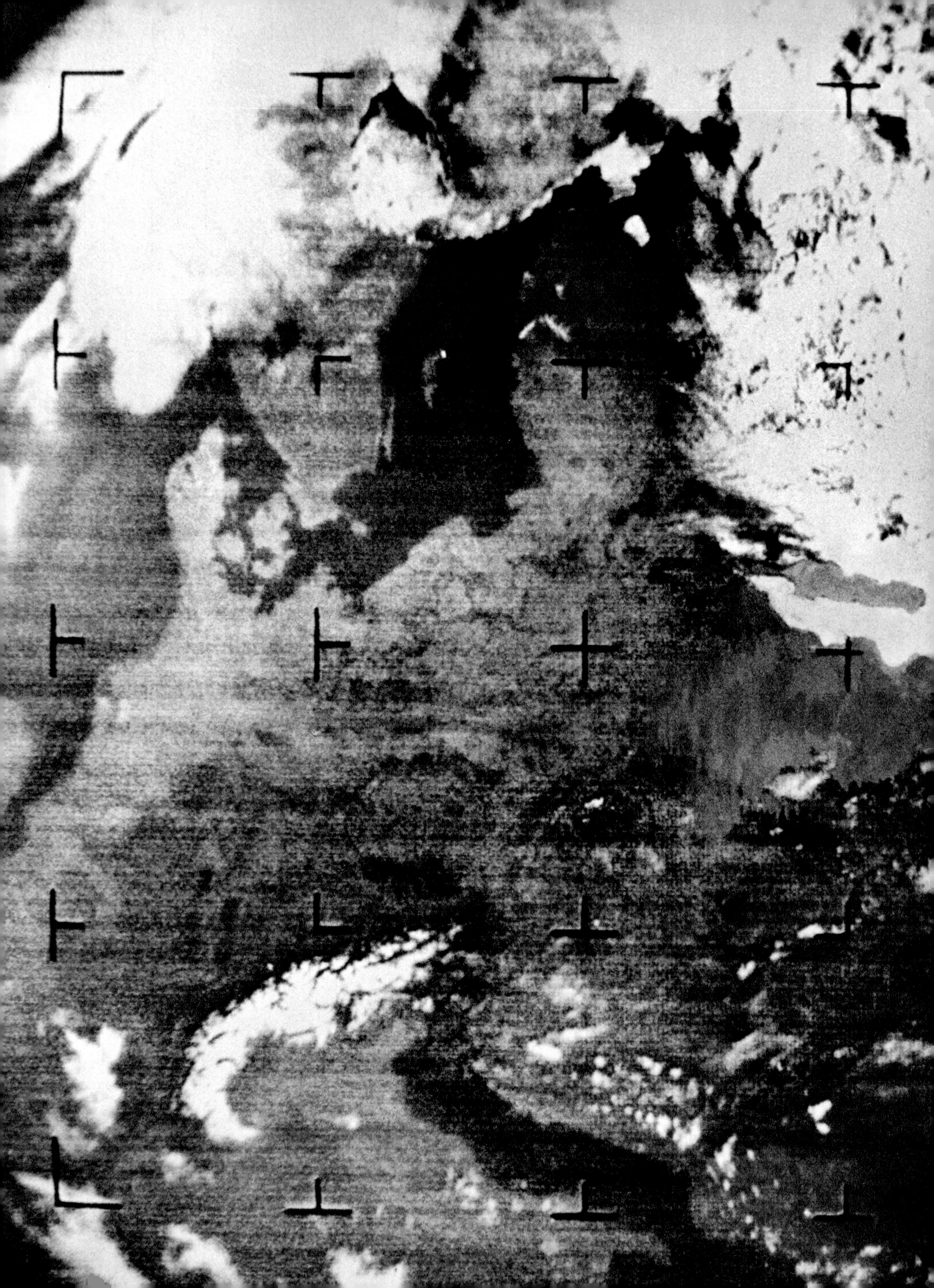

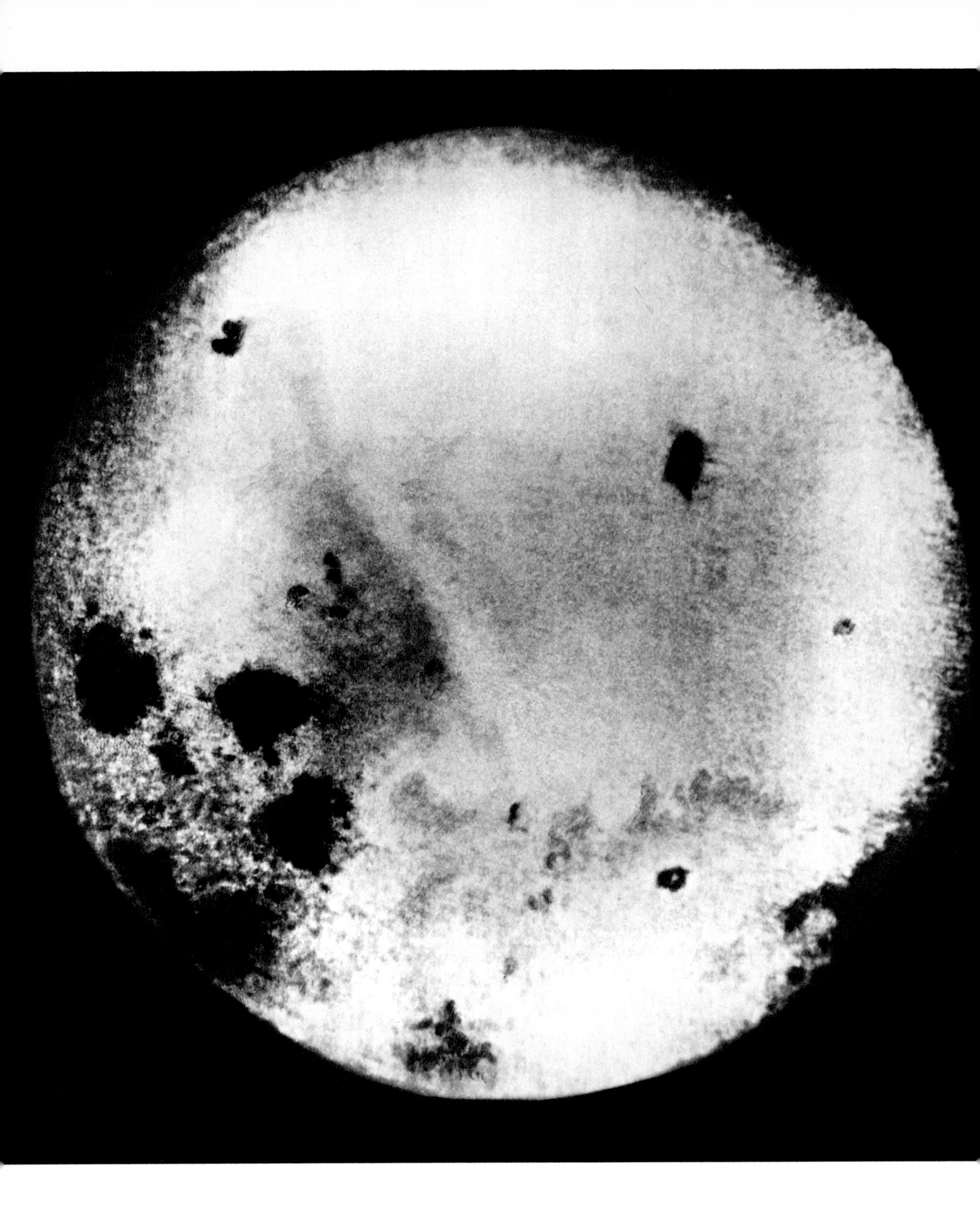

PICTURE CAPTIONS

Please fold out ▸